The Secret Life of GENES

The Secret Life of GENES

Decoding the blueprint of life

Derek Harvey

FIREFLY BOOKS

A FIREFLY BOOK

Published by Firefly Books Ltd. 2019
Copyright © 2019 Octopus Publishing Group Ltd
Text copyright © 2019 Derek Harvey

First printing

ISBN-13: 978-0-2281-0175-8

Library of Congress Control Number: 2018963775

Library and Archives Canada Cataloguing in
Publication is available from Library and Archives
Canada

Published in the United States by
Firefly Books (U.S.) Inc.
P.O. Box 1338, Ellicott Station
Buffalo, New York 14205

Published in Canada by
Firefly Books Ltd.
50 Staples Avenue, Unit 1
Richmond Hill, Ontario L4B 0A7

Edited and designed by Tall Tree Limited

Printed and bound in China

Contents

Introduction

It is a mind-boggling fact that all the information needed to make a human being is contained in a microscopic speck: a fertilized egg just a fraction of the size of a full stop on this page. The speck consumes nutrients to grow bigger, but nothing is added to the instructions it carries as it develops from a tiny ball of cytoplasm into a living, walking, thinking person. It's all there from the very start. The instructions themselves are held in special threads of DNA, some of which (if you could tease them out) are as long as your thumbnail. But they are so slender and bundled up so perfectly that you can't usually see them, even down a microscope. When this single-celled fertilized egg divides into two, the threads package up into solid structures called chromosomes. Only then can we actually see our genetic material, and even then only if we stain the chromosomes with special dye and magnify them a few hundred times.

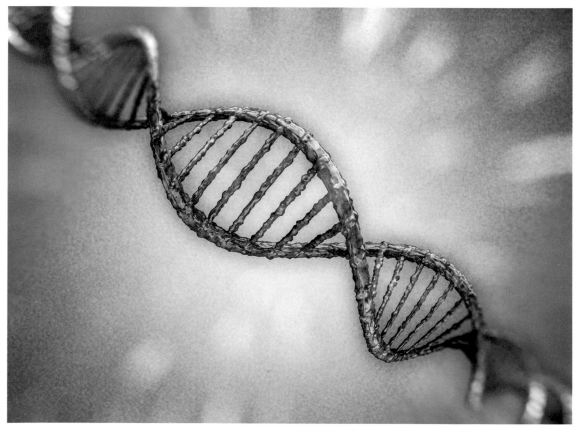

The complex double-helix shape of a DNA molecule carries the genetic blueprint from which a living being is made and is responsible for creating differences between individuals.

The Genes That Make Us What We Are

Here, in this microscopic world, is the key to everything that makes us human. Pieces of DNA called genes (20,687 of them in humans, according to the latest count) carry discrete bits of information, each with a very specific purpose. One gene instructs the skin to make the substance collagen, which keeps it firm and elastic; others help to make pigment, or instruct our cells on how to generate energy. About five or six weeks into a pregnancy, genes kick in to make us male or female. There are even genes that make chemicals in our brains that will affect our behaviour long after we are born. As we grow older, outside influences also conspire to make us what we are: bodies become fat or thin depending on the food we eat, and our brain stores memories that change how we behave. But our ability to process food or to think very much depends upon our genes.

So where did all our genes come from? In the minutes leading up to our conception, we did not exist but our genes already did. Half of them were bundled in the head of a swimming sperm, and half in an egg waiting in our mother's fallopian tube. Only when these two extraordinary cells joined was our own unique genetic identity born. Unless, in a moment of spontaneity, this fertilized fleck splits to make a genetically identical twin, we come into this world genetically unique. This book will help us to understand why there has never been another you, and there never will be. We begin by looking at exactly what genes are, and how they are constructed of atoms arranged in very particular ways. Then we can see how they act to make life grow and reproduce.

Genes Through Deep Time

Of course, other living things also have genes, and our genetic heritage did not begin with our parents: it goes back to the very origin of life on Earth, more than 3 billion years ago. The sperm and eggs that carry genes from one generation to the next are themselves a product of deep ancestry. Genes can be traced back to our grandparents, our great grandparents, and all the way back to our prehistoric ancestors that first abandoned the trees for walking upright. Genes are passed down in this way because they are self-replicating, meaning that, when a cell divides, it generates copies of its genetic information to make not only more cells, but more humans. But this copying process is not perfect, and the gradual, but inevitable, accumulation of "imperfections" is what makes us all different. Ultimately, it accounts for the evolution of new kinds of life. Given enough time, over millions of years and countless generations, it is therefore hardly surprisingly that, the further back we go, the more different our ancestral genes were. But, astonishingly, despite the passage of time, some or our genes have changed very little. There are genes in our bodies today that are so similar to genes in single-celled microbes that the only reasonable explanation is that they have been conserved that way from some distant common ancestor. If it seems unthinkable that we have evolved from something so simple across billions of years, just think back to our own humble origins inside the womb, where a comparable process produced a recognizably human body from a single cell in just nine months. Genes really do carry the secret of life through the ages, whether this is in a human lifetime or across earth-time itself. The second part of this book moves on from the behaviour of genes inside cells to how genetic heritage and variation come together to account for the biggest biological journey of all: evolution.

Chemical Genes

Despite the remarkable things that they can do, genes are still at root chemical substances. In 1828, a German chemist called Friedrich Wohler discovered that urea, the waste chemical substance found in our pee, could be made in the laboratory. For the first time, it was shown that substances produced in the living body obeyed ordinary rules of chemistry: they contained no extra mysterious life-giving spark. Just decades later, the material of genetics was found to be no exception. DNA, short for deoxyribonucleic acid, was found to be one of the chemicals of life, and, as neatly demonstrated by a pea-growing monk called Gregor Mendel, its component genes are transmitted as particles from generation to generation in recognizable patterns that we call inheritance. A century later, DNA was revealed to be molecule with the shape of the twisted ladder: the iconic double helix. Since this discovery, scientists have found ways to analyse the genetic differences between people and other organisms, and even to manipulate genes to create new kind of life forms. By the turn of the millennium, researchers had even succeeded with the unthinkable: cataloguing our entire genetic makeup in the Human Genome Project. In the new age of genetic engineering, such chemical techniques are performed routinely today and even have the potential to cure inherited diseases. As we shall see in the final chapters of this book, we are living at a time when geneticists are making some of the most exciting advances in science.

After four weeks, the single fertilized egg inside the womb has developed into an embryo. Still only about the size of a poppy seed, it is already starting to take a human shape, each stage in its development controlled by its genes.

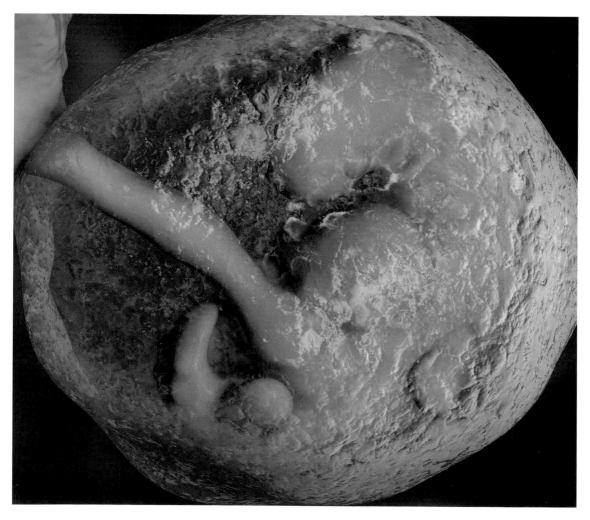

Chapter 1
SECRETS IN THE BLOODLINES

How Does Inheritance Work?

For a long time, the true nature of genes was a mystery. It took the ingenuity and persistence of 19th-century scientists to track it down, a journey that helped them solve one of life's greatest mysteries: inheritance.

Ancestry and inheritance are so integral to the story of life that we take them for granted. Children take after their human parents just as oak trees grow from the acorns of other oak trees. An Alsatian dog, pregnant by another Alsatian dog, will only produce Alsatian pups, and even microbes can pass on qualities that make some of them give us disease or make others resist our attempts to control them with drugs. In every case, something

"in the blood" (or "sap" or cytoplasm) must be making living things the way they are: something that gets passed down, generation after generation, whenever they reproduce. But this secret something presides over impressive changes too. A week-old human embryo – just a microscopic hollow ball of cells – can grow and develop into a woman or a man. And creatures that roamed the Earth millions, or even billions,

A dog will inherit features from its mother and father. The rules of ancestry mean that a puppy with two Alsatian parents will grow to be an Alsatian.

Secrets in the Bloodlines

of years ago were very different from the ones alive today. So how can the secret in the bloodline keep Alsatians much the same, but also turn a ball of cells into a person, and make primitive crawling microbes evolve into walking beasts?

Legacy from the parents

Inheritance was one of the last great mysteries of biology to be solved by scientists. Long after brilliant minds worked out how the heart pumped blood and how the stomach digested food, scientists were still pretty clueless when it came to reproduction and ancestry. Today, we know that animals and plants are made up of billions or trillions of cells, and that a body starts off as a microscopic single-celled speck: a fertilized egg. The information needed to build the body has to be in that speck. From it develop limbs, a heart, stomach and brain, or leaves and roots and flowers. How can this be possible?

With the invention of the first microscopes, 17th century scientists convinced themselves that they could see tiny creatures inside sex cells. Most saw them in sperm: within the heads of each tailed cell, they saw a tiny huddled foetus. The idea was that life was pre-formed, with each generation waiting to break free, one inside the other, like Russian dolls, going all the way back to Adam. Once inside the woman's womb, the foetus was nourished and simply grew and grew.

But there were obvious flaws with this idea. Quite apart from the logistical problem of cramming every generation past – and presumably future – into every sperm cell, if life came pre-formed from the father's loins, why did offspring inherit characteristics from both parents? The woman's contribution had to be more than just nourishment. And why did some babies turn out to be male and others female? In the end, the theory collapsed in the 19th century when better, more powerful microscopes proved that these pre-formed minibeasts were simply not there. Scientists had discovered that the maternal contribution consisted of a microscopic egg, and that sperm and egg had to join at fertilization to create a baby.

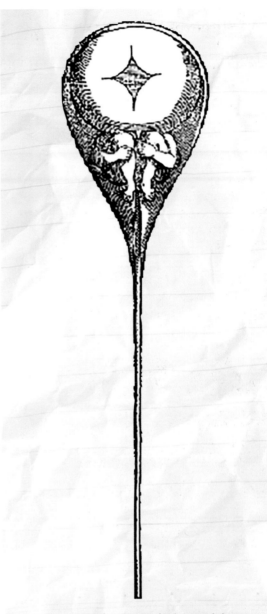

This illustration from 1695 depicts a miniature person inside a sperm. A theory called preformation held that each sperm or egg contained a homunculus ("little person") that would grow into the next generation.

Whatever was going on in the mysterious field of inheritance, it seemed to be passing through microscopic sex cells from both parents.

tantalizingly close to the solution when he wrote that heredity came down to the transmission of a message, like a blueprint. But no-one seemed to be taken with Aristotle's insight.

Darwin himself made a breakthrough in the story of life with his theory on evolution, but was less successful when it came to inheritance. He imagined that every part of a parent's body released special chemical factors (he called them gemmules) that circulated in the blood and ended up in the sex organs to pass into sperm or eggs. These factors affected the offspring, depending upon where they had been produced: those from the eyes governed the colour of the eyes, those from the long bones affected height, and so on. The factors coming

The Greek philosopher Aristotle (384–322 BCE) suggested that male sperm contained something that he called a "formal cause" that shaped the female's substance during fertilization. Many have likened his idea to a genetic message.

Bloodlines Through the Blood

In the 19th century, biologists reverted to an earlier idea that some sort of special life-giving essence flowed through sperm, eggs or, most likely, both. By the time of Charles Darwin, a popular idea was that this essence came from all parts of the body and converged into the sex organs. The essences from male and female then combined at fertilization. This was, at least, closer to the truth than preformed foetuses. Back in Ancient Greece, the philosopher Aristotle had even come

Chopping off a mouse's tail may annoy the mouse, but it makes no difference to the tails of its offspring.

Secrets in the Bloodlines

from father and mother perfectly blended when a sperm fertilized an egg, so the next generation carried a mixture of characteristics from each parent. The idea of a blood-like bloodline sounded tempting, but it was still wrong.

The biggest problem was that this kind of blending inheritance would surely always produce intermediates. Just like mixing red and blue paint to make purple, blending bloodlines would dilute biological variety until living things would all end up looking the same. Life is not like this. The theory was to become a particular embarrassment for Darwin because it patently contradicted his idea that evolution could amplify differences and even create new species. Also, some inherited characteristics, such as red hair in humans or albinism in mice, can skip a generation only to reappear in later descendants. Any theory of inheritance had to explain this, too. Blending inheritance could not. And what if body parts were altered in life, such as by accident? Would this affect their chemical factors that get passed down? A German biologist called August Weismann proved that this was not the case. He cut off the tails of five generations of mice, but all the new mice stubbornly developed tails of normal length.

The answer to the inheritance problem finally came from an unexpected source: a monk who was growing peas in a monastery garden in Brno, then part of the Austrian Empire. Gregor Mendel's painstaking breeding experiments and skill with statistical mathematics would give him the scientific breakthrough needed and eventually earn him the title of "Father of Genetics". The trouble was, no-one at the time noticed what he had done.

Inheritance by Particles

Mendel shunned the microscope in favour of counting pea plants. By breeding, and crossing, varieties of garden peas, he concluded that each variety was caused by something discrete that was passed down intact, generation after generation. He had made the critical discovery that inheritance was down to particles, not "paint". When pure-breeding purple-flowered peas were crossed with white-flowered ones, only purple flowers came through in the first generation of offspring. But white ones emerged in the generation after that. The colours never merged, and he deduced that a particle causing white flowers must have been temporarily hidden in the first generation, only to emerge further down the line. Inheritance involved a process more akin to mixing beads of different colours than stirring paints. Although the beads mingled, red and blue beads stayed red and blue and never blended into purple. Inherited traits depended on the way the beads were shuffled. Under the right circumstances, blue might be "hidden" by red, or might re-emerge, just as characteristics can skip a generation. Today, we call these particles genes, and Mendel's theory of particulate inheritance was eventually recognized as one of the most important breakthroughs in the history of biology.

Gregor Mendel (1822–84) wrote that he wanted his pea plant studies to address the "history of the evolution of organic forms". He discovered that inheritance depended upon particles that today we call genes.

According to the theory of blending inheritance, purple-flower pea plants crossed with white-flowered ones would produced offspring with intermediate colour: pale purple. But Mendel found that the hybrids all had purple flowers. The white-flower characteristic had disappeared.

Pure-breeding purple-flowered pea plant

Pure-breeding white-flowered pea plant

Offspring produced if inheritance worked by perfect blending.

Offspring actually produced by Mendel's experiments

It wasn't until the turn of the 20th century, 15 years after he had died, that Mendel got the recognition that he deserved. By then, biologists in Holland and Britain had "rediscovered" his account of the pea plant experiments and realized they were the key to the mystery of mysteries. Mendel's work was replicated and lauded, as a new branch of gene-related science, dubbed genetics, was born.

Particles of Change

Genes might be conserved from generation to generation, but today we know that they can control biological change too. As a fertilized egg grows into an adult body, the genes in its cells get copied from one cell to another so that all the cells end up genetically the same. But genes get switched on or off throughout the development process. This means that different parts of the body are formed in the right places at the right time, all under the control of different sets of genes. And over many generations, across great stretches of evolutionary time, genes themselves mutate from one form to another. But details of the exact ways in which genes are involved in the complex processes of development and evolution would take genetics much further than Mendel would have ever imagined.

If Mendel's hybrid plants were inter-bred, blending inheritance would predict that the generation after that would still all have pale purple flowers. In fact, Mendel found that the white-flowered plants reappeared, making up about a quarter of the offspring.

Hybrid purple-flowered pea plant

Hybrid purple-flowered pea plant

Hybrid purple-flowered pea plant

Hybrid purple-flowered pea plant

Offspring produced if inheritance worked by perfect blending

Offspring actually produced by Mendel's experiments

Genetic Blueprints and Recipes

The genes in a living thing work as sets of instructions. They contain the information needed to build the living, working body, and, therefore, determine its characteristics.

Today, we know that every kind of living thing contains thousands of genes, each one contributing in some way to the overall organism. Sometimes, as with Mendel's peas, there is a simple one-to-one relationship between a gene and a clear, identifiable, characteristic. For instance, there is a gene that determines the colour of a pea plant's flower, another that influences a pea plant's height at maturity, and so on. But in most cases – something that Mendel was not to know – the relationship is more complex than this, with each gene having multiple effects. Genes are often

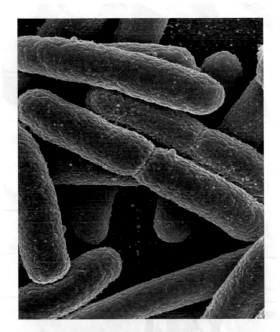

Escherichia coli **bacteria, such as these from the human gut, may be tiny, but they still contain about 5,000 genes.**

likened to blueprints, but taken together they are really more like recipes. A blueprint shows the positions of all the "nuts and bolts" for what is being made. You could even, if you wish, recreate the blueprint from the thing that is made. Living things aren't built like that. Each gene has a discrete role to play, but many gene effects get blurred as they work together to build an organism, just like the ingredients involved in making a cake. For instance, contrary to a simple version you may have been taught in school, there could be more than a dozen different genes that affect human eye colour. In fact, it was pure luck that Mendel chose an organism where genes were straightforward enough that the laws of inheritance were revealed so beautifully. The simplest kinds of organisms, such as common bacteria, are controlled by some 5,000 genes; humans have about 20,000. And all life forms contain their genes in one or more microscopic building blocks called cells.

Genetic Information

Bacteria and many other microbes are single-celled, but an adult human body consists of around 60 trillion cells that make up all the body parts. And every cell carries the genes needed to build that particular body. This means that every cell of your body (with a few exceptions, such as mature red blood cells, which have dispensed with their genes), carries the instructions that could potentially make another you. The total amount of information in your entire body is therefore phenomenal. One estimates equates it at around 40 zettabytes. That's 40 million times more information than in all the different books and other printed materials ever published.

In reality, the total genetic information in all the cells of your body, barring the odd copying mistakes that happen when you grow, is practically the same. Genetic variety comes with the genes in different bodies and especially between bodies of different species.

Communities of Genes

Practically all the complex life we see around us – people, birds, trees – is built on two sets of genes. The 20,000 different kinds of genes needed to make a human are actually doubled up inside each human body cell, so two sets – 40,000 genes in total – are crowded inside that microscopic space. Fruit flies have 14,000, doubled up to 28,000. The sets are not exactly the same. Each kind of gene can come in different varieties, called alleles. In pea plants, for instance, there are two alleles of the flower colour gene: one allele instructs for purple flowers, the other instructs for white. Sometimes, like this, there are two alternatives, but many genes come in alleles of lots more different types. The existence of alleles is the reason we display so much variation. When two different alleles find themselves together in the same cell, the outcome depends on the way the alleles interact. The purple-flower allele in garden peas dominates the white allele. In a similar way,

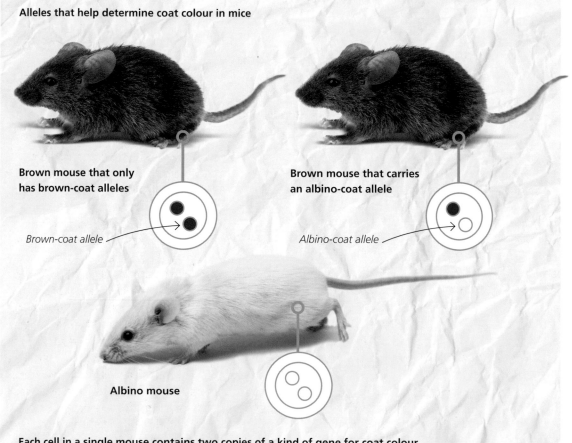

Alleles that help determine coat colour in mice

Brown mouse that only has brown-coat alleles

Brown-coat allele

Brown mouse that carries an albino-coat allele

Albino-coat allele

Albino mouse

Each cell in a single mouse contains two copies of a kind of gene for coat colour. The commonest variety (allele) of this gene produces ordinary brown-grey coat, but a second variety makes it albino. Both copies must be the albino allele for the mouse to be white.

Secrets in the Bloodlines

Each sex cell, whether it's a sperm or an ovum, carries its own set of genetic information.

an albino allele in animals can be masked by an allele that causes dark pigment.

The reason for organisms having two sets of genes in this way comes down to two points. Firstly, it helps to have a backup set in case any one gene goes wrong. The right genes – the instructions for life itself – are critical for a proper, healthy working body. But if one, say, is miscopied as the body grows, it is possible that its healthy backup allele will take over and override the deficiency of the error-ridden allele. As we shall see, this system is not infallible, but it's still better than pinning all your biological hopes on a single set. The second reason for having double sets of genes has more long-term implications. It comes down to sex.

Genes and Sexual Reproduction

Sexual reproduction helps to mix genes up. Offspring are genetic mixtures compiled from two parents, and every time a sperm fertilizes an egg a wholly new set of genes, with new combinations of alleles, comes into being. This means that in a sexually-reproducing population, each genetically different individual is differently equipped to face the world. Sex improves the chances of someone doing well. A population of genetically identical clones could be exterminated by a new disease, to which they are all equally susceptible. But there is a potential problem with sexual reproduction. Fusing body cells together during sex would mean continually doubling the number of genes with each union, something that would quickly become unmanageable. Instead, sexual organisms first separate their two gene sets when they form special kinds of sex cells: sperm or eggs. When a cell in a male's testis divides to form sperm, one set of genes goes to one sperm and one set goes to another. The same thing happens in a female's ovaries when she makes eggs. Each sperm or egg carries a single set of genes (and so half the complete baggage of genes of a normal body cell), but still has a complete set of instructions to stay alive. But when sex cells join at fertilization, the double set is restored to make the next generation, ready to make sex cells of their own.

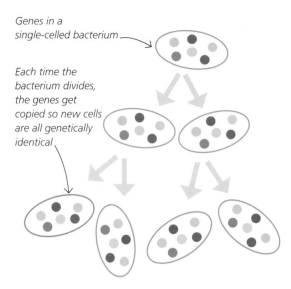

Genes in a single-celled bacterium

Each time the bacterium divides, the genes get copied so new cells are all genetically identical

Genes in microbes, such as bacteria, are copied when their cells reproduce. This means that one ancestor can produce a population of genetically identical individuals. Many kinds of plants and some animals can reproduce asexually in a similar way.

Cells that contain a double set of genes are described as diploid; single-set cells are haploid. All life that reproduces sexually must have a life cycle that alternates between the diploid and haploid states in this way: the sets are separated to form sex cells and united at fertilization. In flowering plants, the haploid condition is passed through pollen, rather than sperm, but the result is much the same. Some life cycles are anomalous. For instance, moss plants are haploid and only become diploid when they sprout their spore capsules. And honeybees have a peculiar system whereby females are diploid and males are haploid. But, even with these complications, the separation and union of gene sets help to mix alleles in their sexual life cycles.

United by Genes

Despite the genetic differences between organisms, a crucial thread of similarity runs through everything alive. Humans share some genes with dogs, cabbages and even bacteria. These universal genes instruct on matters that

are critical for all life, such as how to get energy from food. That we are not the same as a baboon, a banana or a bacterium is due to differences in many genes, but lots of genes are similar too. A remarkable achievement of science since Mendel has been to show that, in spite of all our differences, the underlying secrets of how genes work are remarkably similar, even for things as superficially unalike as microbes and people. This story makes up a big part of this book.

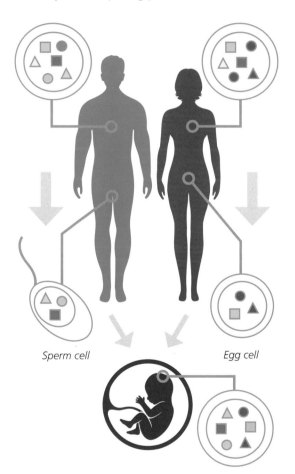

Sperm cell

Egg cell

There are two sets of genes inside each of the trillions of cells of human body. These sets separate when body cells make sperm or eggs, so these sex cells contain half the number of genes as body cells. A set of genes from father and mother mix together at fertilization, creating a foetus made up of cells contains 50% paternal and 50% maternal genes.

Secrets in the Bloodlines

Chapter 2
SECRETS IN THE GENES

Finding the Double Helix

Genes are made of DNA, a molecule that famously has the shape of a twisted ladder. This double helix is key to understanding how genes work, and even to explaining the very nature of inheritance and reproduction.

The body of a living thing is made of organic matter. This means that, except for the water, it consists largely of complex molecules containing carbon. Some of these "molecules of life" are the biggest known to scientists. The molecules of genes, perhaps somewhat unsurprisingly, are among them. But exactly what kinds of molecules are genes?

Scientists knew there were two likely candidates for the "stuff of inheritance". The first is protein.

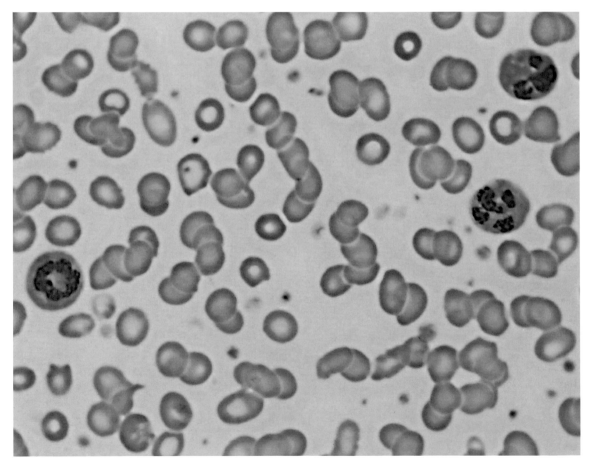

There are two kinds of blood cells. Red blood cells (the small discs) predominate, but they have lost their genetic material during their development. Like other cells of the body, white blood cells contain DNA in their nuclei, here stained dark purple.

Protein is everywhere in a living body. It comes in many different forms, but all proteins contain the same basic ingredients of carbon, hydrogen, oxygen, sulphur and nitrogen atoms. Proteins are abundant components in cells – they predominate in muscle and blood – and were prime candidates for controlling the bloodline. But protein had a rival called nucleic acid. Mysterious nucleic acids, containing the same elements as protein, are found packed deep inside each cell but it was not obvious what function they performed. Some scientists thought they were some sort of scaffolding for supporting the critical protein. But others had a hunch that they were more important than that.

Nucleic Acids

In 1869, shortly after Mendel had published the results of his pea plant-breeding experiments, the first steps towards understanding the chemistry of genes were being made at the University of Tubingen in Germany, 700 km (450 miles) from Mendel's garden. Here, a physician called Friedrich Miescher was interested in finding out more about the nucleus. The nucleus was the "heart" of a cell and looked as if it should contain its secret control centre. Miescher was not especially focussed on the mystery of inheritance, but wanted to tackle the cell's nucleus from a chemical viewpoint: he wanted to know about the stuff of its ingredients. But studying the chemical makeup of tiny cells, and particularly the make-up of a part of the cell, has always been a tricky thing to do. First, he needed to get enough cells to work with, and he found them in the suppurating, infected wounds of patients at the nearby hospital. Pus is rich in infection-fighting white blood cells, and each cell has a very clear nucleus. He washed the cells off pus-soaked bandages, managed to separate their nuclei and successfully isolated an entirely new kind of nuclear substance. He called it nuclein.

Over the following years, other scientists worked hard to reveal the true identity of Miescher's nuclein. They refined his methods to obtain purer samples and discovered that the

As well as being the first person to isolate nucleic acid, Friedrich Miescher (1844–1895) was also an early advocate that material inside cell nuclei was the basis for heredity and the passing on of features from one generation to the next.

key ingredient was acidic, so they called the nuclein nucleic acid. Then, in the early decades of the 20th century, an American biochemist called Phoebus Levene discovered that there were actually two kinds of nucleic acid that differed somewhat in their chemical make-up. One was called ribonucleic acid (RNA), and the other, because it contained less oxygen, deoxyribonucleic acid (DNA). But mystery continued to surround the function of nucleic acids. Although scientists were understanding more and more about their chemical composition, no-one yet knew what they did.

A crinkly version of pneumonia bacteria is destroyed by the body's immune system, but the smooth version is lethal. Experiments showed that a chemical substance from dead lethal bacteria could flow into the benign one, turning it dangerous. This substance was found to be DNA.

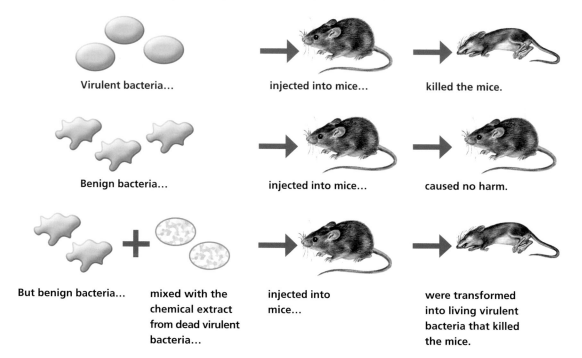

Virulent bacteria…   injected into mice…   killed the mice.

Benign bacteria…   injected into mice…   caused no harm.

But benign bacteria…   mixed with the chemical extract from dead virulent bacteria…   injected into mice…   were transformed into living virulent bacteria that killed the mice.

Which Chemical is in the Gene?

Clever experiments showed that molecules of DNA, like many "molecules of life" including proteins, were made up of smaller inter-locked building blocks. Chemists discovered that the chain-like structure of proteins made proteins not only big and complex, but also versatile. Proteins perform many tasks in the body, from driving reactions as catalysts, to building muscle or carrying oxygen (as the red blood pigment protein, haemoglobin). It was for this reason that so many scientists thought that genes were made of protein. Nucleic acids such as DNA seemed to lack the sophistication necessary to do anything important. They even had fewer kinds of building blocks: where proteins had 20 different sorts, DNA had only four, called nucleotides.

The first evidence that DNA was indeed the key to inheritance came in 1928 from the research of a physician working at the British Ministry of Health. Prompted by the global Spanish Flu epidemic that followed World War I, Frederick Griffith was studying pneumonia bacteria. He discovered that a dangerously virulent strain of the bacteria could somehow help transform a benign strain, making the harmless microbe turn lethal. Nothing more than a chemical extract from the virulent bacteria was needed to do it. If the exact ingredient – the "transforming principle" – within this extract could be identified, could this be the chemical secret of the gene? Another decade passed before the answer was found: biochemists not only confirmed Griffith's results, but also uncovered the secret identity. It was, indeed, made of DNA.

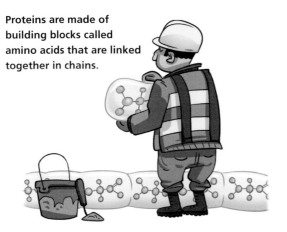

Proteins are made of building blocks called amino acids that are linked together in chains.

Building the Double Helix

Although many scientists were not convinced by the studies of pneumonia bacteria, others were taken with the idea that DNA was more than some sort of support for the cell's control centre, even though it apparently had a simple, regular form and lacked the elaborate configurations of proteins. One technique, in particular, was consistently pointing to a simple DNA shape. X-ray diffraction is a way of analysing the shapes of molecules by studying the way they scatter X-rays onto photographic film. Each molecular shape produces a particular pattern of illuminated spots, and scientists skilled in the technique can use the pattern on the photograph to work out the shape of the molecule that produced it. DNA produces a characteristic cross-like pattern, and this told scientists it was shaped like a helix.

X-ray diffraction had been developed in the 1950s. Maurice Wilkins and Rosalind Franklin, working at King's College, London, pioneered the study of DNA in this way. They inspired an uneasy collaboration with ambitious researchers at Cambridge University, James Watson and Francis Crick. Using materials in the university's workshop, Watson and Crick built a model of DNA using the X-ray diffraction data, and established that DNA was actually made up of two chains of nucleotides that were entwined around each other: the famous double helix. The critical information-carrying parts of its nucleotides, called bases, pointed inwards and joined to form the "rungs" of the long twisted ladder.

In 1953, James Watson (left) and Francis Crick (right) pooled all the available information about DNA and built a model of its structure at Cambridge University. They had deduced it was a coiled two-stranded molecule: the double helix.

The Secret in the Bases

There are four kinds of bases in DNA: two bigger ones, called adenine and guanine, and two smaller ones called thymine and cytosine. Watson and Crick deduced that the bases were paired in the rungs of the DNA ladder in a particular way. Not only is one big base always opposite a smaller one (to keep the same distance between the sides), but adenine is always paired with thymine, and guanine with cytosine.

The sequence of bases along the DNA chains is key to the inheritance of characteristics: the sequence is "read" by the cell and used to determine an organism's features. But the special base pairing also means that the sequence along one chain of the double helix automatically determines the sequence along the other: the two base sequences fit together like the parts of a jigsaw puzzle.

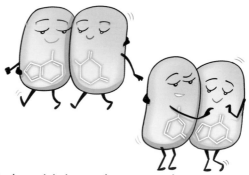

The bases join in complementary pairs, with adenine twinned with thymine and guanine twinned with cytosine.

This, in itself, has little to do with the way genes govern inherited features: in reality, only one chain matters. But, as we shall see in chapter 5, it had enormous implications for understanding how living things reproduced.

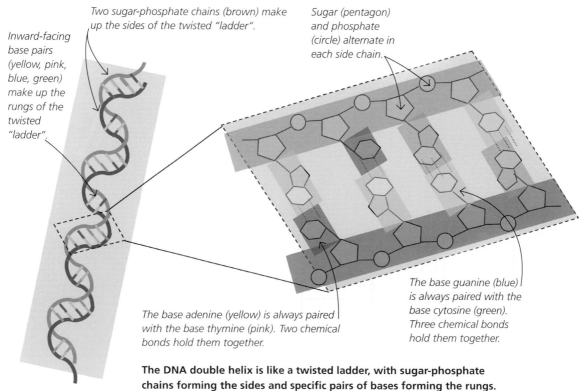

Two sugar-phosphate chains (brown) make up the sides of the twisted "ladder".

Inward-facing base pairs (yellow, pink, blue, green) make up the rungs of the twisted "ladder".

Sugar (pentagon) and phosphate (circle) alternate in each side chain.

The base adenine (yellow) is always paired with the base thymine (pink). Two chemical bonds hold them together.

The base guanine (blue) is always paired with the base cytosine (green). Three chemical bonds hold them together.

The DNA double helix is like a twisted ladder, with sugar-phosphate chains forming the sides and specific pairs of bases forming the rungs. The instructions in a gene are held in the specific sequence of bases along one side of the ladder.

How Genes are Arranged

Genes, the particles of inheritance, are sections of DNA double helix. Each gene spans a particular sequence of DNA bases. Longer molecules of DNA can accommodate more genes.

Genes are made of DNA, but what, exactly, is a gene? Mendel discovered that genes were inherited as particles that could shuffle around in different combinations as they passed from generation to generation. Does this mean that a single gene corresponds with a single DNA double helix? No. A cell is controlled by thousands of genes, but only has a modest number of DNA molecules.

From DNA to Genes

Each different kind of organism needs exactly the right sort of genes to make and maintain its body. Bacteria need several thousand; more complex plants and animals need tens of thousands. But cells do not contain thousands of molecules of DNA. Instead, many genes are linked together on the same DNA double helix. A gene corresponds to a particular section of DNA, a stretch of bases running along one of the two twisted strands, that encodes for a protein. The protein's activity then helps to contribute to the body's inherited characteristics. All genes are found on one of the two strands, although different genes are found on different strands.

By the standard of other molecules, a single DNA double helix is very long. On average, one double helix from a human cell, spanning millions of bases, is about 5 cm (2 in) long, an astonishing length for something that fits into a cell just a hundredth of a millimetre across. The recipe for a body has a fixed number of genes, and each species of living thing carries a fixed number of DNA molecules, or chromosomes. Human body cells contain 23 different molecules of DNA, and human genes are scattered at fixed positions on each one. For instance, a gene that helps to give us blue eyes is about a quarter the way along the length of DNA molecule number 15. (The molecules are usually numbered in size order, from biggest to smallest.) Each kind of gene has a fixed locus, or address, within the DNA. But body cells contain two sets of genes, so there are 23 pairs of molecules in a human body cell, with two "doses" of number 15, and two "doses" of this eye colour gene. Although the loci of the two doses of the gene are identical, they could have different versions, or alleles, of the gene. For instance, one dose could be for blue eyes, and the other for brown eyes. During sexual reproduction, these two sets of genetic material separate, so sperm and egg cells carry only one set of each.

Two doses of a gene at a fixed locus (position) on a DNA molecule

More precisely, each gene corresponds to a length of one strand of the double helix.

Two DNA molecules in a cell are highlighted here. These molecules both carry the same sorts of gene at the same positions.

Other kinds of living things have their own distinct arrangement of genes and DNA molecules. Chimpanzees have a total of 48 DNA molecules per cell, houseflies have 12 and Mendel's garden peas have 14. But the number is not related to our perceived notion of complexity. There is a small butterfly called an Atlas blue which has 452 DNA molecules in its cells. Clearly, it's not the size of the DNA that typically matters too much; it's what the cell does with it.

Non-coding DNA

One of the most surprising things to emerge from detailed studies of DNA within the last 50 years is the fact that genes are connected by stretches of DNA that seem to be functionless. In fact, depending upon species, genes might account for less than 5 per cent of an organism's DNA. In humans the fraction could be as little as 2 per cent. Genes along DNA are often likened to the beads on a necklace, but non-coding DNA makes this comparison fanciful. In reality, genes are separated by long gaps. And not only do genes vary in size, but the gaps do, too. Again, the lengths involved – both for genes and "gaps" – depend upon species. Of course, the gaps are only gaps in encoded information. The "gaps" are still made up of bases, it's just that the bases there do not encode for characteristics. If you could run your eye along a base sequence of an entire length of a double helix (a feat that would be something like scanning 5,000 pages filled with continuous four-point text), all you would see would be a continuous stream of letters (AGCACTG…), with no punctuation and no clear indication where each gene began and ended.

Why are the information gaps there at all? The short answer is that no-one knows the exact reason. The non-coding DNA has traditionally been called "junk" DNA, to reflect its lack of function. Some scientists think that it represents accumulated genetic "baggage" that has built up over millions of years of evolution: DNA that once had protein-encoding function, but is now redundant. But it is possible that much so-called "junk" DNA really does have a role to play, but that its secret has yet to be cracked. Some DNA, for instance, plays a part in engaging with other parts of the cell to help move genetic material about during cell division.

Almost 98 per cent of a human's DNA is non-coding and has no effect on how a human appears and functions.

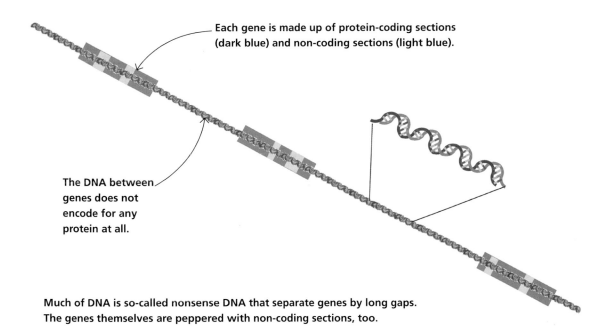

Each gene is made up of protein-coding sections (dark blue) and non-coding sections (light blue).

The DNA between genes does not encode for any protein at all.

Much of DNA is so-called nonsense DNA that separate genes by long gaps. The genes themselves are peppered with non-coding sections, too.

Perhaps more surprising than the information gaps between genes is the fact that gaps occur within them too. These interruptions to the protein-encoding message are called introns. (The encoding, so-called expressed, regions of a gene are called exons.) Introns have roles to play in controlling and regulating genes. Some act like switches, so they can trigger the activation of a gene when it is needed, or turn it off when it is not. This helps to explain how different genes can only affect different parts of the body. Because the introns do not contain information explicitly needed to make protein, when genes are "read" to form proteins (see chapter four), the introns are edited out, so only the information held by the encoding bits of DNA is actually used.

The Genetic Difference Between Bacteria and More Complex Cells

Bacteria are the simplest kinds of living organisms. They are made up of single cells, and each cell, itself, is smaller and structurally simpler than the cells that make up bodies of animals and plants. Notably, the DNA of animals and plants is packaged inside a capsule called a nucleus, like a cell within a cell. Bacteria, instead, have loose DNA bundled in the open cytoplasm.

There are other differences. The ends of a bacterial DNA double helix are joined up, so the entire molecule is like a ring. More complex cells of animals and plants have open-ended linear threads of DNA. More complex cells typically contain around ten times more genes than bacteria: a single important case where size of the genetic package really does matter. The extra genetic baggage in animals and plants is a reflection of greater size and complexity. More genes are needed to provide the extra instructions needed to make a multi-celled body, and to keep the cells working properly together. But extra DNA can be a problem when it comes to cell division: how can so much DNA by copied and accurately partitioned into new, growing cells? Animals and plants solve it by using a special packaging trick: they form chromosomes.

Chromosomes and Karyotypes

Every time a cell divides, its DNA must be copied and properly distributed to the new cells. The loose DNA risks becoming entangled when this happens, so the cell bundles it into tighter packages called chromosomes.

If you look at a growing cell down an ordinary school microscope, its DNA appears unremarkable. You certainly won't see a double helix or genes. You could stain it with special dyes, but you would still only see a coloured blob. But when cells divide, something extraordinary happens. Each molecule of DNA becomes shorter and thicker as it coils more tightly around itself: each is transformed into a solid thread that can be seen with appropriate magnification. These threads are called chromosomes. Chromosomes appear only when the cell is dividing, and are needed to stop the long double helices getting tangled up. Cells divide as a body grows. And, in preparation for this, all the DNA molecules, and so all their genes, have to be copied. During division itself, the copies must be exactly distributed to the two new cells, so each receives the same critical set of genes. This amazing feat is achieved, despite the fact that all the original DNA molecules, which are centimetres long, are inside a space that is a tiny fraction of the size of a full stop. But by packaging each one into a shorter, neater chromosome, the DNA can be shuffled into position without problem.

Chromosomes do not form in bacteria (although bacterial DNA is sometimes, misleadingly, called "bacterial chromosome"). This is because they are not needed. The simple loops of DNA in bacteria can be copied and distributed without the risk of entanglement.

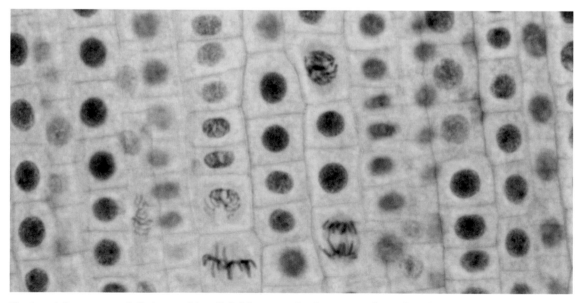

Plant root tips are especially busy with cell division to make them grow through soil. These onion root cells show how blobs of DNA, stained blue, transform into spidery chromosomes as cells prepare to split.

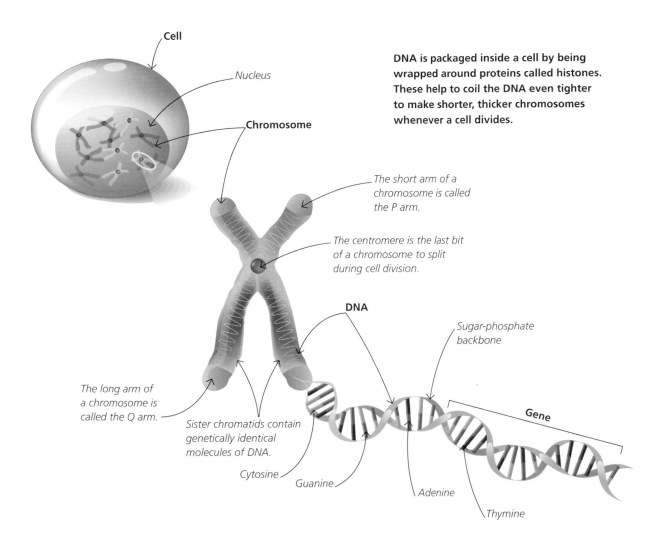

Cell

Nucleus

Chromosome

DNA is packaged inside a cell by being wrapped around proteins called histones. These help to coil the DNA even tighter to make shorter, thicker chromosomes whenever a cell divides.

The short arm of a chromosome is called the P arm.

The centromere is the last bit of a chromosome to split during cell division.

DNA

Sugar-phosphate backbone

The long arm of a chromosome is called the Q arm.

Gene

Sister chromatids contain genetically identical molecules of DNA.

Cytosine

Guanine

Adenine

Thymine

How Chromosomes are Formed

DNA is not alone in making chromosomes. Inside every cell there are special proteins that cling to the DNA to give it support. These proteins form a kind of scaffold, with the DNA draped around them. (This is somewhat ironic, because, as we saw in the earlier section, scientists first thought that these roles were reversed.) When a chromosome is made, the proteins help to make each DNA double helix coil more, and then coil again. By doing this, the coil gets thicker, but, more importantly, gets shorter too. In fact, it coils so much that a couple of centimetres of DNA contract down to a fraction of a millimetre. This

makes it far easier for the cell to move all its DNA during cell division. When chromosomes appear, the normal workings of each gene temporarily stop. They start up again when cell division is finished and chromosomes unravel back into the longer, invisible threads inside the new cells.

Because chromosomes can be seen so clearly with the right magnification, they have helped scientists to appreciate just how much DNA is in cells, and how its arrangement varies in different kinds of living things. Each visible chromosome corresponds to a separate double helix with its own set of linked genes.

Karyotypes

If cells are caught at just the right stage of cell division, their migrating chromosomes are nicely separated. And when these chromosomes are stained, photographed and enlarged, scientists can not only count them, but even arrange them into their pairs. What is more, a special stain, called Giesma stain, gives each different chromosome a characteristic purple banding pattern. The darkest bands appear wherever the DNA contains a lot of A and T bases (adenine and thymine). These bands do not mark out the position of genes, but are useful marking points for mapping out the length of the chromosome, and for distinguishing one chromosome from another.

This arrangement of stained chromosomes, either shown as a photograph or a diagram, is called a karyotype. Each species of organism has its unique karyotype, with a specific chromosome number, chromosome structure and banding pattern. Comparing karyotypes can reveal clues about species relationships. For instance, the 46-chromosome karyotype of a human is very similar to the 48-chromosome karyotype of a chimpanzee. Alternatively, karyotypes can reveal genetic disorders, such as human Down's syndrome.

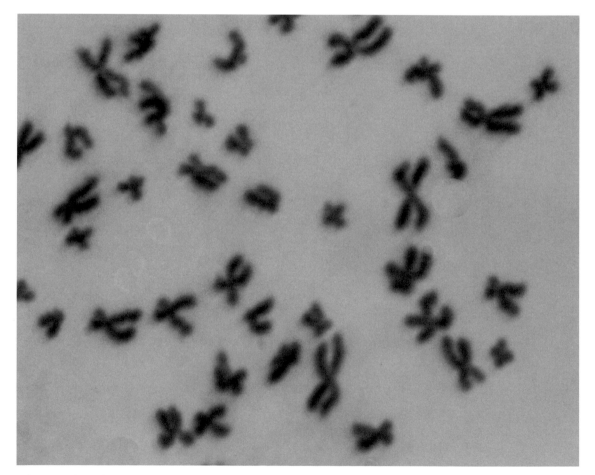

A human white blood cell caught in the middle of cell division shows stained chromosomes that are nicely separated. Photographs like these can be enlarged and the images of the chromosomes matched up into their pairs, ordered by size, to make a karyotype.

Chromosome Structure

Chromosomes in a karyotype vary not only in size and banding pattern, but also in shape. By the time chromosomes become visible, each one contains two replicas of their DNA molecules, and towards the end of cell division these replicas will tease apart before separating entirely to go to separate daughter cells. A constriction point somewhere along the chromosome, called a centromere, is the last to split. Some chromosomes have their centromere near their mid-point. Others have them towards the end. This is just another property that geneticists can use to help match chromosomes up in their pairs.

A cell copies all its DNA before dividing, so each chromosome has two genetically identical DNA molecules. When the centromere splits towards the end of cell division, one DNA molecule will go to one new cell and one to the other.

Sex Chromosomes

Sex in some organisms, such as certain kinds of plants, is down to genes that are scattered throughout the chromosomes. This means that the karyotype alone gives no information about sex. In some animals, sex even depends upon the surroundings. For instance, turtle eggs incubated at higher temperatures hatch into females, while those incubated at lower temperatures hatch into males. Genes that control sex are switched on or off according to temperature. But for many organisms, including humans, there is a sex-specific pair of chromosomes.

In humans, and other mammals, one pair of chromosomes (conventionally depicted last in a karyotype) contains the genes that determine whether an individual develops into a male or female. In mammals, one of these sex chromosomes – the larger of this pair – is called the X chromosome; the smaller of the pair, indeed the smallest in the entire karyotype, is called Y. (These terms are applied due to historical reasons and have nothing to do with the chromosome shapes.) Individuals that inherit two copies of the X chromosome develop into females. Those that get a copy each, of X and Y become males. In birds, the sex chromosomes are labelled W and Z, and females have two different sex chromosomes (WZ), while males have the same (ZZ).

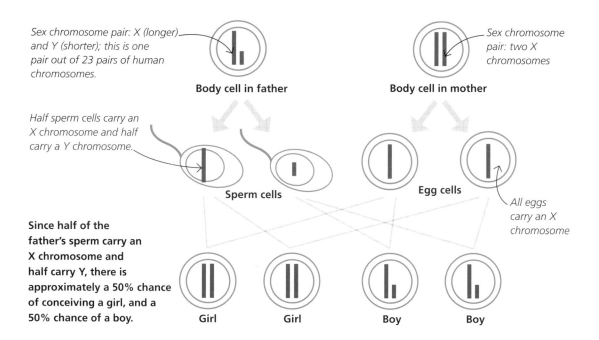

Sex chromosome pair: X (longer) and Y (shorter); this is one pair out of 23 pairs of human chromosomes.

Body cell in father

Sex chromosome pair: two X chromosomes

Body cell in mother

Half sperm cells carry an X chromosome and half carry a Y chromosome.

Sperm cells

Egg cells

All eggs carry an X chromosome

Since half of the father's sperm carry an X chromosome and half carry Y, there is approximately a 50% chance of conceiving a girl, and a 50% chance of a boy.

Girl   **Girl**   **Boy**   **Boy**

When sex is caused by different chromosome complements, one has two different sex chromosomes and the other has sex chromosomes that are alike. And the sex with different sex chromosomes always determines the sex of the next generation. In humans, and other mammals, the father carries this distinction. The sex chromosomes separate whenever sperm or eggs are made, just like all the other chromosome pairs in the body. This means half the father's sperm carry an X chromosome, and half carry a Y. All the mother's eggs carry an X. If an X-loaded sperm fertilizes the egg, the result is a girl (XX), while a Y-loaded sperm produces a boy (XY).

Chromosome Errors

Studies of karyotypes help to diagnose genetic disorders that arise when chromosomes are not properly rearranged after cell division. As we have just been reminded, when sperm or eggs are made in the sex organs, the cells involved must separate their double sets of genes so that each sex cell receives just a single set. This is achieved by halving the chromosome number. In humans, the double-set of chromosomes in the body cell reduces down to a single set of 23 chromosomes inside each sperm or egg. Pairs of chromosomes become single.

Mistakes can occasionally occur in this halving process. One or more pairs do not separate properly, so both members of a pair end up in one sperm or egg. Its opposite daughter cell now lacks any of the crucial genes, and so dies, but the sex cell carrying an extra chromosome can still engage in fertilization. Down's syndrome, for instance, arises because of a failure involving chromosome pair number 21. When an egg cell with an extra such chromosome is fertilized, will develop into a person with Down's syndrome, where all their cells have three doses of chromosome 21, making 47 chromosomes in total. (Chromosome errors of this kind are explained in more detail in chapter 9.) The problems associated with these genetic conditions show that the correct balance of genes is important for healthy development. While each gene is critical for keeping a body alive, an extra dose can also cause defects.

Chapter 3
WHAT GENES DO

One Gene, One Protein

**Each gene influences a cell, and so the entire body, in a subtle way.
It instructs the cell to make a special kind of protein,
which in turn affects what the cell can do.**

As molecules go, genes are not especially hard-working. Instead, they are a repository for a vast amount of information: the information needed to build and maintain a living body. The genes inside a cell are like a microscopic library, one that contains thousands of volumes containing information for life. Other components of the cell have the job of following their instructions. These are the hard workers, and one group of molecules are particularly important in this respect: the proteins.

We associate proteins with the foodstuffs needed to grow healthy bodies: to build muscle and make us strong. That is true, but the remit of proteins goes much, much further than that. Proteins are the main workhorses of a cell or organism. Some help move substances into and out of cells, and even around the entire body. Others help drive vital reactions to process food, build materials and release energy. In fact, it is no exaggeration to say that practically every biological process can be linked to some sort of protein. Not surprisingly, they have an enormous impact on characteristics. And because proteins are assembled by following the instructions in the library of genes, these characteristics are inherited. Why do some of us inherit brown eyes? The short answer is that a gene encodes for brown eyes. The long answer is that at least eight genes influence eye colour, but one (OCA2) has a particularly big effect. These genes make proteins that shape the manufacture of brown pigment. Each gene instructs the cell to make a protein. This one gene–one protein principle is key to understanding a gene's secrets.

Proteins are some of the hardest-working substances in the body and control thousands of different tasks that happen inside every one of your cells.

Enzymes that are enzymes that drive aspects of metabolism, such as for energy-release, building big molecules, photosynthesis

Proteins that are enzymes that drive aspects of metabolism, such as for energy-release, building big molecules, digestion

Proteins in membranes, such as for absorbing food from the gut, regulating of salt and sugar responding to signals

Proteins in membranes, such as for absorption into roots

Structural proteins needed to strengthen plant cells, help fight infections

Blood proteins that circulate oxygen, help fight infections, help blood to clot, carry signals

Muscle proteins needed for muscle contraction

Structural proteins needed to strengthen the skin, build bones

Thousands of different kinds of proteins are produced to drive the workings of a body, and each type of organism needs its own collection. Every protein is produced using inherited instructions carried by a particular gene.

Reading the Genes

Genes are read continually by a living cell. Proteins work so hard that they eventually wear out, so new ones have to be made to replace them. At any one moment, the thousands of genes inside the body are being scanned by the cell's gene-reading machinery and the body is making the thousands of kinds of proteins according to their instructions. We say that each gene encodes a particular protein. Some proteins are needed in greater abundance than others. Red blood cells are packed with a protein called haemoglobin and contain little else. Haemoglobin is the red blood pigment that carries oxygen around our bodies, and we need a lot of it. In fact, as red blood cells mature, they even lose their DNA to cram as much of the red stuff inside. Each red blood cell lasts for three or four months before it is recycled. As a result, these cells, and their haemoglobin, have to be replaced at a phenomenal rate. They are generated in bone marrow at the speed of 400 million every hour. The haemoglobin-producing cells of the marrow are working overtime when they read the haemoglobin genes. Compared with that, proteins that work as hormones, such as insulin, are produced in smaller quantities. Only a tiny amount of insulin is needed to do its job of controlling sugar levels, but its presence is no less critical.

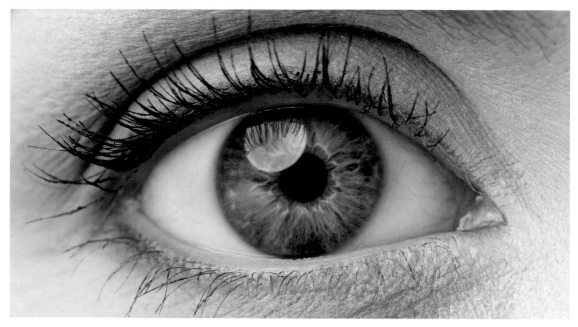

Your eye colour is determined by the amount of melanin in the front layers of your iris. Blue eyes have less melanin than brown eyes. The production of melanin is controlled by several genes, particularly two on chromosome 15.

Remember that, specifically, a gene is a length of a DNA double helix, corresponding to the base sequence along one of its chains. Because in the intact double helix, these bases are bonded to the complementary bases on the opposite chain, the double helix must unravel whenever a gene is read. The cell manages to achieve all this, even though it has to read thousands of genes at a time. This also helps to make clear why the entire gene-reading, protein-making process must temporarily stop whenever a mature cell splits in two during cell division.

How Genes and Proteins Affect Characteristics

When we think of inheritance, we think of the obvious characteristics that get passed down through the generations: things like human eye colour, blood group or the flower colour of garden peas. But most inherited characteristics are passed down in a less straightforward way. Human height does not come in discrete categories. Instead, height comes in a range of values, with most people clustered around the average. And some characteristics are less than obvious, but still vital for life. All living things carry out seven basic processes: nutrition, respiration, excretion, sensitivity, movement, growth and reproduction. The way each kind of organism performs each function must be inherited too. These are the invisible, but crucial, elements of inheritance, because life itself depends upon them.

Working proteins are at the heart of all these factors but most attributes depend upon hundreds, and sometimes even thousands of different kinds of proteins that work together in a coordinated manner. Many, called enzymes, drive chemical reactions.

Enzymes are needed to keep life's processes ticking over and each process demands a different enzyme. Respiration alone relies on dozens of different enzymes that process sugar and fats stage-by-stage to release usable energy. And instructing the manufacture of all these proteins are the genes in each cell.

What Genes Do

Different Proteins for Different Body Parts

The body of an organism is very complex, with each part concerned with a particular task. The heart has the job of pumping blood. Blood, itself, transports food, waste and chemical signals. A limb is obviously very different from a brain. However, each cell in the body carries the same sets of genes that have been copied from the original fertilized egg. When sperm met egg, the resulting combination of genes was a mixture from mother and father. But after that, as the fertilized egg divided into two, then into four, eight, and so on, all of the genes were copied to supply the same information for each new generation of cells. But clearly genes exert their effects differently in different parts of the body. Even though the haemoglobin gene is found in every cell with a nucleus around the body, it only instructs the manufacture of haemoglobin in red blood cells. Everywhere else, the gene is silent.

In fact, genes do not blindly churn out proteins wherever they occur. Genes respond to cues and triggers from their surroundings. These triggers help to ensure that some genes are switched on, while others are switched off. This system is crucial for explaining how a fertilized egg can grow into an adult body, with lots of different working parts. Even at that very early stage, the cells that result from those first divisions carry slight chemical differences that mean some of them a destined to become skin and others produce a nervous system.

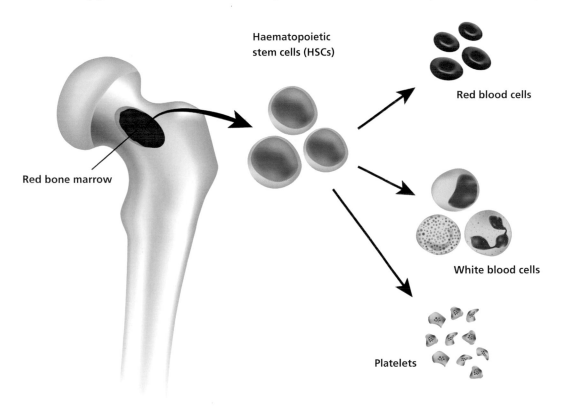

Haematopoietic stem cells (HSCs)

Red blood cells

Red bone marrow

White blood cells

Platelets

Blood cells are made inside the bone marrow from special cells called haematopoietic stem cells. Several types of cell are made, depending on how protein-encoding genes are activated. Cells destined to become red blood cells use genes for making oxygen-carrying haemoglobin, whereas white blood cells use genes for making proteins that fight infection, such as antibodies.

Genes and Development

The way in which a single cell can grow into a body with all its different working parts was, like genetics itself, one of the great mysteries of biology. Even after genes had been discovered, found to be made of DNA, and found to encode for proteins, still no-one really understood how an embryo developed. But the same gene–protein system that governed the workings of an adult body eventually explained the secret of the embryo.

The basic arrangement of an embryo's head and tail ends is determined as early as the fertilized egg. Proteins made in the mother's body (and, therefore,

instructed by her own genes) spread through the egg and, according to their distribution and concentration, determine the embryo's front and back. This simple start affects the genes carried by the egg, and all the subsequent embryonic cells that spring from it. A particular level of protein, for instance, switches on "front-acting" genes that will form the head. Different levels switch on "back-acting" genes for the tail. As the embryo gets bigger and more complex, more complicated patterns of proteins impact different parts of the growing body so that genes in some places are activated to form skin or limbs, and so on.

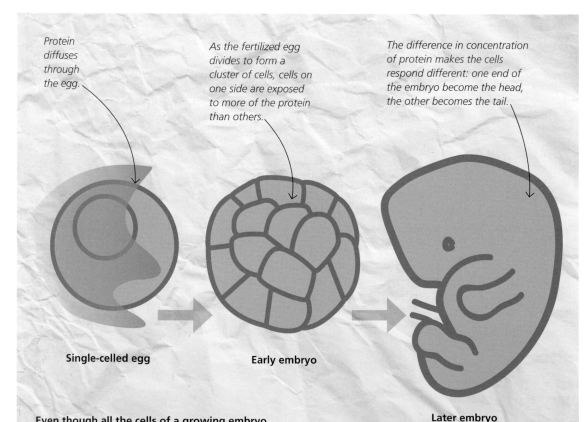

Protein diffuses through the egg.

As the fertilized egg divides to form a cluster of cells, cells on one side are exposed to more of the protein than others.

The difference in concentration of protein makes the cells respond different: one end of the embryo become the head, the other becomes the tail.

Single-celled egg

Early embryo

Later embryo

Even though all the cells of a growing embryo contain the same genes, they are exposed to different levels of maternal protein. This ensures that genes are switched on in different parts of the embryo to make it develop different parts.

How Proteins Work

Thousands of kinds of proteins are made by the thousands of different genes. Each one performs a vital task inside the body and, collectively, they determine the characteristics of an organism.

The most abundant kind of protein in the world is probably one made only in plants. This is the protein that grabs carbon dioxide (CO_2) from the air around us so that other enzymes can turn it into sugar. It is called RuBisCO (ribulose bisphosphate carboxylase). There is far more RuBisCO in the world than all the haemoglobin in all the animal bodies put together. Part of this is because the world contains such a weight of green leaves, places where this remarkable food-making process, called photosynthesis, takes place.

But it also stems from what RuBisCO is actually doing. Turning carbon dioxide into sugar is difficult. Indeed, getting carbon dioxide to chemically react in any way at all is little short of miraculous. RuBisCO has to work hard to persuade carbon dioxide to engage with the food-making machinery inside the leaf cells. This is why plants need so much of it. But practically all life depends upon the food chain created by vegetation's photosynthesis, meaning that RuBisCo, and the gene that encodes it, is probably the most important gene–protein system in the history of life on Earth.

Most kinds of proteins in the world fall into the same class as RuBisCO: they are known as enzymes. These are proteins that catalyse specific chemical reactions inside living bodies. This means that they help drive substances to react. Indeed, in the absence of such a catalyst, they would not do so. Not all enzymes have such a tough time of it as RuBisCO, but most are also needed to keep organisms alive.

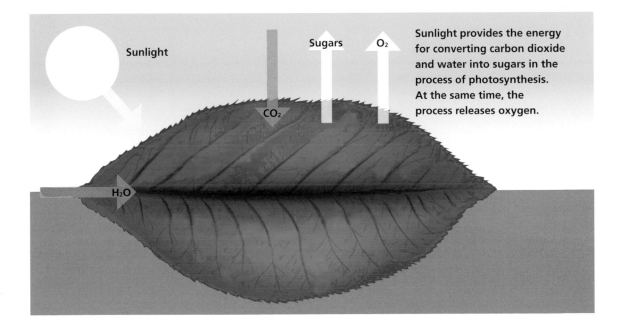

Sunlight · Sugars · O_2 · CO_2 · H_2O

Sunlight provides the energy for converting carbon dioxide and water into sugars in the process of photosynthesis. At the same time, the process releases oxygen.

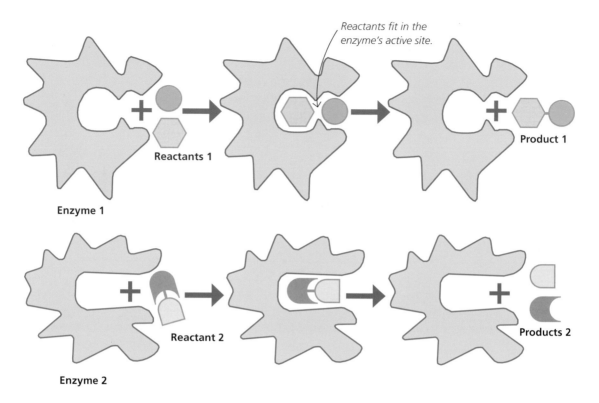

Reactants fit in the enzyme's active site.

Reactants 1

Enzyme 1

Product 1

Reactant 2

Enzyme 2

Products 2

Different enzymes perform different tasks because they have different shapes. Each enzyme shape accommodates a particular shape of chemical reactant into an active site, where the chemical reaction takes place to form products.

The Secret of Enzymes

Keeping a body alive involves a vast array of chemical reactions. Collectively, these reactions are called the metabolism. Every organism has a busy metabolism, from the lowliest bacteria to plants and animals. The processes include reactions that assemble big molecules from small ones, helping the body to grow and repair, as well as countless others that break molecules down or change one kind of substance into another. Virtually none of these reactions would be successful outside the body, or, if it were, it would work too slowly to be effective. Enzymes encourage molecules to react at a higher speed appropriate for life, and to do so in very precise ways. Each different reaction of the metabolism needs a particular sort of enzyme to catalyse it.

Enzyme molecules are specific because of their shape and chemical properties, which are ultimately determined by instructions in genes. These special shapes help enzymes to lock onto other molecules of complementary configurations, so they fit together like a lock and key. The bonding is only temporary, but it is enough for the enzymes to bring their targets together in just the right way to ignite a chemical reaction. The different shapes of other enzymes, encoded by other genes, mean that they latch onto other kinds of targets, and so catalyse different kinds of metabolic processes. Just as RuBisCO can only drive a reaction in photosynthesis, so amylase, an enzyme in our spit, will only drive the digestive breakdown of starches in food.

A Multitude of Tasks

Enzymes make up more than half of all the known kinds of proteins. So what about the rest? In fact, they too rely on their special configurations to do their proper jobs. Haemoglobin has the right shape for holding iron, which, in turn, binds to oxygen. Muscle proteins include long fibres that interlink and slide against one another to make a muscle contract. The fibrous protein keratin is found in the skin, hair and nails. Other proteins may be equally familiar, at least by name. Infection-fighting antibodies are protein that bind to, and neutralize, potentially harmful invaders. And many (but not all) hormones are proteins. Parts of the body produce hormones to communicate with other areas. Insulin, for instance, binds to the liver when our body is flooded with sugar from a high-carb meal, and triggers the liver cells to store up the excess.

Most hormones can communicate in a specific way because their targets have special surface receptors, and these are proteins too. In this way, by encoding for proteins, genes have an impact on practically every working part of the living body.

This receptor protein is bound to a cell's membrane.

This hormone protein circulates in the blood.

When the hormone binds to the receptor, it triggers the cell to respond.

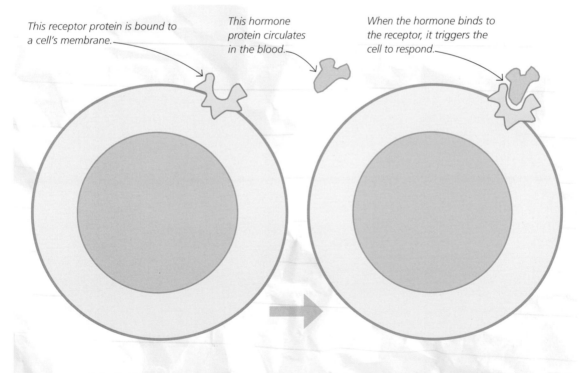

Many events in the body involve more than one kind of protein working together. Hormones are chemical messengers that are produced in one part of the body and effect a response in others. Complementary receptors ensure they only affect the right cells.

Controlling the Cell Cycle

When cells have matured, they may split to make two cells from one. With each round of cell division, the cell's genes must be copied to provide a complete set of instructions for each new cell. This happens when the body is growing, and is also necessary to replace cells that wear out and die. Some cells divide faster than others. Bone marrow cells involved in producing blood cells divide prolifically, but many cells in the brain scarcely divide at all in the mature body. Bacteria are among the fastest, able to split on average once every 20 minutes, but most human body cells are only able to split every 24 hours.

The speed of this cell cycle depends on many factors. Some of these are purely environmental, so most bacteria divide more quickly if conditions are warmer. Cell division also responds to chemical cues, such as substances called growth factors. Their presence can trigger a cell to replicate its DNA and divide. And, at the heart of it all, the cell has a special clock-watching system that ensures all the events of the cell cycle progress in the right order.

Control of the cell cycle is especially important in a multi-cellular body. This is because the growth and number of cells must be balanced and coordinated in different areas. If one part grows too quickly, it could invade and swamp others: the consequence would be a cancerous tumour.

Proteins, and the genes that encode for them, are intimately linked to every part of this cell cycle. Many of the growth factors that flow around the body and bind to cells to trigger cell division are proteins, and there are other proteins inside each cell that communicate the action of the trigger deep into the cell. At the same time, there is a special family of proteins, called the cyclins, that rise and fall during the cell cycle. The peak of each kind of cyclin stimulates a particular process, beginning with copying the DNA and finishing with splitting the cell. The cyclins work by activating other proteins, such as the enzymes needed to drive critical chemical reactions. There are even genes and proteins that trigger certain cells to self-destruct: a process called apoptosis. These cells have to be sacrificed so that the body can develop properly. It happens, for instance, to eliminate the webbing that exists between the digits of a foetal hand or foot, to release separate fingers and toes. Altogether, taking into account the growth factors, the triggers, the cyclins and the enzymes, there are dozens of kinds of protein involved in controlling the cell cycle, and each one is encoded by its own gene.

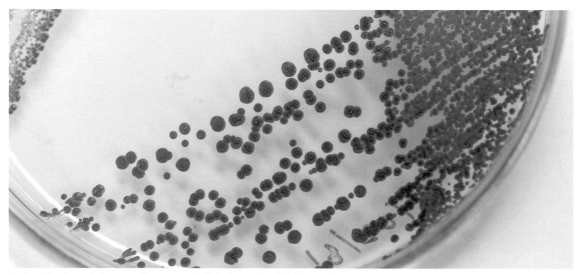

Under ideal conditions, such as a petri dish, bacteria divide quickly to produce visible colonies.

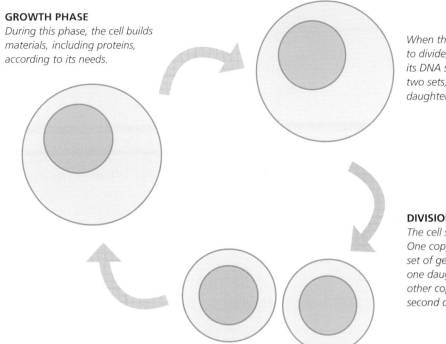

GROWTH PHASE

During this phase, the cell builds materials, including proteins, according to its needs.

When the cell is ready to divide, it copies all its DNA so there are two sets, one for each daughter cell.

DIVISION PHASE

The cell splits in two. One copy of its entire set of genes goes to one daughter cell; the other copy goes to the second daughter cell.

The cell cycle describes the life of a cell through the stages of its growth and cell division. Each stage of the cycle is precisely controlled by a set of proteins, helping to ensure that tissues grow at an appropriate rate.

Proteins for Making Proteins

A central idea in biology is that there is a one-way flow of information, from genes to proteins. In other words, genes instruct the assembly of proteins, and not the other way around. But genes are not entirely independent of their protein products. Like all the other complex molecules in the living body, the stuff of genes, DNA, needs to engage in reactions, too. When DNA self-copies before cell division, a complex set of reactions is called for to build the replicas. "Reading" the genes and making proteins involves still more chemical reactions. These processes must be catalysed by yet another battalion of enzymes. Some proteins, therefore, stay close to the site of their birth in the cytoplasm to help drive the very reactions that produced them.

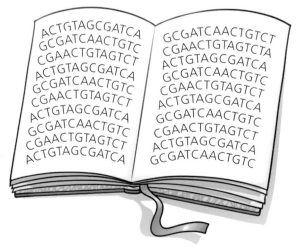

During replication, each gene sequence is read before it is then copied and duplicated.

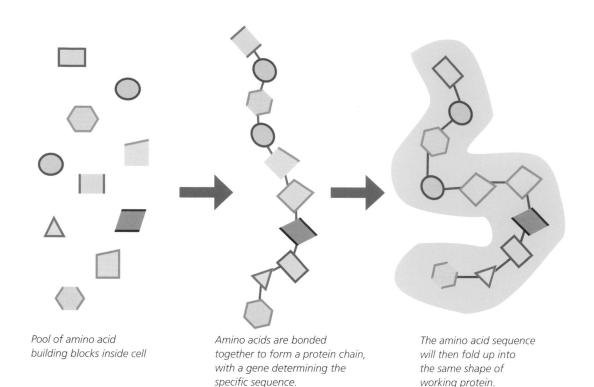

Pool of amino acid building blocks inside cell

Amino acids are bonded together to form a protein chain, with a gene determining the specific sequence.

The amino acid sequence will then fold up into the same shape of working protein.

Each different kind of protein contains a unique sequences of its amino acid building blocks. The order of amino acids along its chain determines how the chain folds. Each sequence produces a unique shape.

Why Proteins Are So Versatile

Proteins can come in so many different shapes and sizes because of the variety of their building blocks. When a cell makes a protein, it draws upon a pool of 20 different kinds of building blocks called amino acids. The message carried by a gene in the form of its base sequence is interpreted so that the cell links together specific kinds of amino acids in the right order to make a protein chain. In effect, the base-sequence "language" of the gene is translated into the amino acid sequence of the protein. The process is so efficient that a cell can generate many hundreds, if not thousands, of chains of exactly the right type from one gene every minute. But loose protein chains are too shapeless to work. They need to fold into their correct configuration.

Fortunately, the cell doesn't need to do much more to get its precious protein shapes. Once a chain of amino acids is assembled, it must fold up into a complex three-dimensional arrangement. There are chemical factors inside the cell, including other proteins, that help this to happen, but overall the particular sequence of amino acids will always fold up in the same way to make the same shape. This happens because each kind of amino acid building block has a unique chemical property. For instance, some amino acids might carry a positive charge and be attracted to ones with negative charge, or repelled by others that are positive. In this way, the base sequence of a gene not only determines the amino acid sequence of a protein, but also its final shape, its chemical properties, and so what it can do.

When Genes Go Wrong

With so much at stake, it is unsurprising that cells do whatever they can to keep the gene–protein system working close to perfect. But errors can happen, and when they do the effects can range from the mild to the catastrophic.

Genes vary a great deal depending upon the instructions they carry and, therefore, the proteins they encode. A gene that encodes for a digestive gut enzyme has a very different base sequence compared with one that encodes for an enzyme that is used in a plant's photosynthesis. But individual genes, which are always found in their gene-specific fixed position (locus) on a DNA molecule of chromosome, can have subtle variations too. The variants, called alleles, may differ by just a few bases and are a natural basis for the variation that exists in a population of one kind of organism. They continually arise by random errors of replication: a process called mutation. Humans vary in skin colour because of allele variations in pigment-encoding genes.

Even slight changes to a gene's base sequence has an effect on its encoding message, leading to a difference in the amino acid sequence of its encoded protein, and therefore the protein's shape. Because a protein's shape is so intimately linked to what it can do, gene changes are most likely to impede a protein's function, or may even stop it from working altogether. Less commonly, the protein's new shape may actually help it perform better, or even do something completely different. For instance, a variant allele of a gene encoding for an enzyme called tyrosinase prevents this enzyme from making melanin, the dark pigment found in many animals. This results in an animal that lacks dark pigment: an albino. Some malfunctioning proteins have little or no effect on the welfare of an individual. Others, such as energy-generating enzymes, are more critical, but cells have gene-checking systems in place and usually have a way of preventing the spread of the damage: they abort or self-destruct.

These albino lion cubs have a gene variant that stops them from making melanin, giving them pale skin and white fur.

Red blood cells collect oxygen from air inside the lungs. At the same time they get rid of waste carbon dioxide so that it can be breathed out.

Although the cells' screening process corrects or eliminates most protein-coding errors, a tiny proportion get through, and are harmful enough to be recognized as diseases. These base sequence errors persist and wreak havoc on the living body.

Disorders of the Blood

Haemoglobin is the protein in the blood that makes it red. Each molecule of haemoglobin contains four proteins chains of two kinds, and so needs two kinds of gene to produce it. Each chain grasps a nugget of iron. It is the iron that allows haemoglobin to perform its job of carrying oxygen around the body, picking it up from the air breathed into the lungs, and off-loading it to all around the body.

The cells need oxygen for their respiration, using it to extract energy from food. But haemoglobin

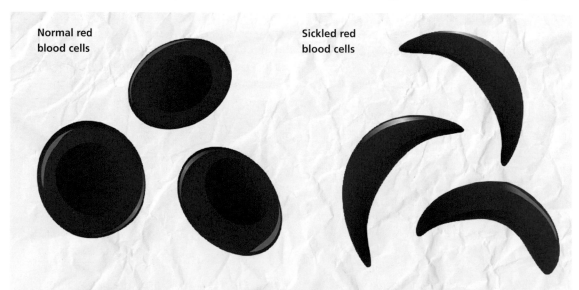

Normal red blood cells

Sickled red blood cells

Just a tiny change in the base sequence of the haemoglobin gene, where the base thymine is replaced by the base adenine, will substitute the wrong amino acid in the protein chain, and cause sickle cell anaemia. The faulty chain folds up the wrong way inside the haemoglobin molecules, making them clump together inside red blood cells, giving them a sickle shape.

What Genes Do

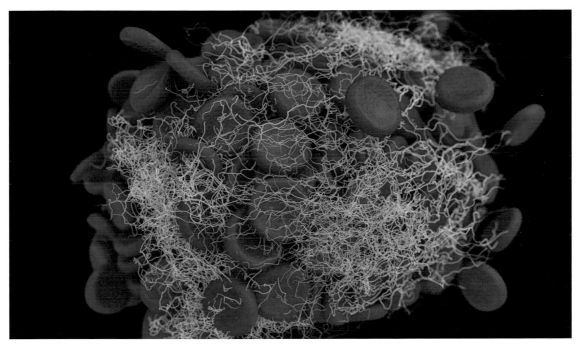

In a blood clot, a meshwork of protein fibres helps to trap the blood cells. Anything that interrupts this process can result in haemophilia.

needs just the right shape to package inside red blood cells and discharge this task properly. Just one wrong amino acid in one of its chains is enough to upset the balance. Sickle cell anaemia results from one such change. At position 6 on one kind of protein chain, the amino acid glutamic acid appears as valine instead. It throws haemoglobin's otherwise perfect shape awry, making the body's red blood cells assume an awkward sickle-like shape. These sickle-shaped cells can clog blood vessels, causing painful joints and even strokes.

Another gene error leads to a failure in clotting. Here the situation is more complicated, because an entire family of proteins is involved. Blood clotting depends upon a chemical reaction that changes a protein called fibrinogen into another called fibrin. Fibrinogen is soluble – dissolved in the blood's plasma – so flows freely. But fibrin is made of solid fibres that mesh together to trap blood cells, making a clot. However, dozens of other kinds of protein, including many enzymes, are involved in controlling a cascade of chemical reactions that turns one to the other when blood is exposed to air. Any one of the proteins involved in blood clotting could potentially malfunction, leading to a bleeding disease called haemophilia. Two of the commonest, haemophilia A and haemophilia B, arise from errors in the genes encoding for blood proteins called factor VIII and factor IX respectively.

Disorders of the Membranes

Every cell of the body is enclosed by a thin membrane. Proteins are especially abundant in cell membranes, where they appear in many different kinds. Some are the triggers or switches that help the cell respond to chemical signals swirling around them. Others work like little pumps to move substances into or out of the cell, or like doorways or gates to let other things through. Many of these gateways help to control levels of key substances inside the cell.

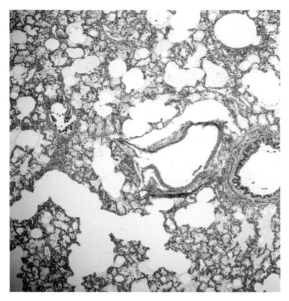

Healthy lung cells under a microscope.

One gateway protein exists in the membranes of many cells that lines internal organs, such as the lungs and the digestive system. It allows salt to leak out from the cell, typically onto a lining that is smeared with mucus. Wherever salt goes, water tends to follow by a process called osmosis. (A similar effect is seen in your finger tips after a long soak. The salts in your skin draw in water, making them go crinkly.) The water then mixes with the mucus, keeping it thin and fluid.

A loss of just one amino acid in the gateway protein (out of a total of 1,480) is enough to cause the disease called cystic fibrosis. It means that not only is the protein folded into the wrong shape, it may not even reach the cell membrane. As a result, salt and water fail to pour from the cell, leaving the mucus thick and sticky, blocking the air ways and obstructing the gut.

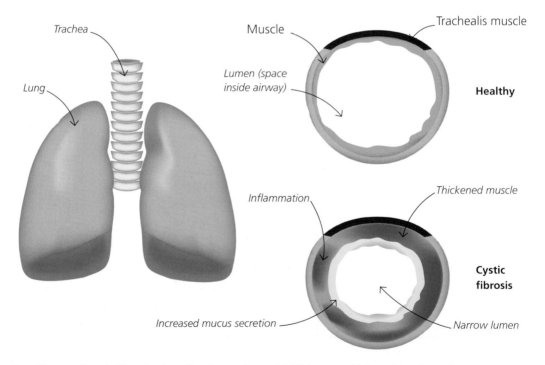

In sufferers of cystic fibrosis, there has been a loss of DNA bases, which means a loss of an amino acid in a regulator protein found in cell membranes. Without it, the protein has the wrong shape for expelling salty water, which leaves a layer of thick, sticky mucus lining tissues.

What Genes Do

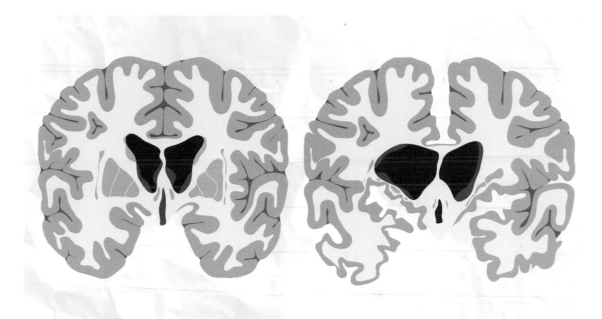

A brain section from a sufferer of Huntington's disease (right), compared with a normal brain (left) shows how brain cells have eroded to enlarge the brain chambers, a condition caused by gene that corrodes a vital brain protein called huntingtin.

Disorders of Muscles and Nerves

One of the biggest genes known in humans encodes for a protein found in muscle called dystrophin, which contains 3,685 amino acids. Dystrophin envelopes muscle cells, helping to anchor the special fibres that are stacked lengthwise inside the tapering cells and make a muscle contract. Dystrophin's chain of amino acids is coiled in places to form springs, which serve as muscular shock absorbers. But its giant gene can be miscopied in several ways, giving rise to a family of progressive muscle-wasting diseases called muscular dystrophy.

Sometimes, diseases can be traced to genes, but the details of what their proteins do are still not fully understood. For instance, a gene that encodes for a particularly large protein (3,144 amino acids long) called huntingtin is needed for the healthy working of brain cells. But scientists do not yet know exactly what huntingtin does, although they know it is critical. There are probably a number of cellular functions, such as signalling

and transporting, that depend on it. And it seems to protect brain cells from the kind of programmed self-destruction (apoptosis) that is so important in the development of other parts of the body. When the protein goes wrong, the result is a condition of progressive brain cell death called Huntington's disease.

When the Cell Cycle Goes Wrong: Cancer

There are so many different kinds of proteins involved in controlling the cell cycle that it is hardly surprising that any underlying malfunctions here can be very complex. But the result can be consistently devastating: cancer. Any disruption to the natural cycle of growth and division of cells could cause cells to grow and spread to invade parts of the body where growth is happening more normally, and a lot more slowly. Over a hundred different kinds of cancer affect humans alone, depending upon the genes and proteins affected and the parts of the body that are involved.

Often as a result of errors in genes that control the natural cell cycle, cancers develop larger numbers of cells with over-sized nuclei. The cells become disorganized as they lose their specialized functions and start to invade neighbouring tissues.

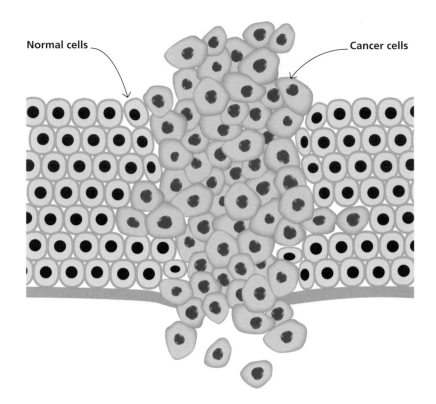

Normal cells

Cancer cells

Genes that control the cell cycle and so have the potential to cause cancer are called proto-oncogenes. (Oncology is the study of cancer.) An error in their base sequence may turn them into cancer-causing oncogenes that make the cells divide invasively to form a growth called a tumour. Typically, cancers result from cells that are triggered to divide when they shouldn't. For instance, proteins that keep cells in check by causing programmed cell death (apoptosis) may malfunction, so cells can escape from their control.

Treating Genetic Disorders

When genetic errors arise in a growing body, only those cells produced after the error will be affected. This is what happens when cancer appears: an oncogene has emerged in a single cell among billions, and a tumour grows from it. As long as the cancer is caught before it is spread too far, then the tumour and its diseased cells can, at least in theory, be removed.

But many other genetic disorders are inherited from the parents and so they are congenital: they are present at birth. Moreover, the gene error was in the original fertilized egg, so every cell of the body produced from that egg will also carry the gene. This means it is impossible to eliminate the error, because the gene pervades every cell and part of the body. Until recently, treatment of these kinds of genetic disorders was restricted to palliative care: helping to relieve the symptoms and so improve the quality of life. However, modern advances in genetic science have the potential to go further than this. Today, it is possible for scientists to change the genetic makeup of cells, for instance by injecting them with working genes to override the faulty ones. This strategy – called targeted gene therapy, whereby healthy genes are used as a drug – has the potential to take treatment much further than ever before thought possible. It is explored in greater detail in chapter 12.

Chapter 4
THE GENETIC CODE

A Code for Life

All living things are built using a code found in each cell. Remarkably, the nature of this code is the same for every species, even though the message it carries can be very different.

A code is a set of rules used to change information from one form into another. Morse code converts a sequence of dots and dashes into a sequence of letters that can be read as a normal language. For instance, "dot-dot-dot, dash-dash-dash, dot-dot-dot" gets changed into "SOS".

The genetic code that governs practically all life on Earth is based on the same sort of rules, but the "language" forms used are the sequences of building blocks in DNA. A cell uses the genetic code to convert the instructions held in a sequence of bases in a gene's DNA into a sequence of amino acids along a protein. Once the protein is assembled, this means that the gene has determined the way the protein chain folds into a unique shape, and so the way it performs a particular task that affects a characteristic. And because genes are passed down through generations, these same characteristics are inherited.

A remarkable discovery made by scientists within the last 70 years is that the genetic code across all life – from bacteria to birch trees and bumble bees, people to parsnips – is almost exactly the same. Although the particular message conveyed can be very different, the code itself is not. This means that a particular DNA base sequence is translated into the same amino acid sequence, whether it is in a microbe, a plant or an animal. This fact has mind-blowing consequences. It is inconceivable that such a complex system could have arisen independently in all those species. The conclusion is therefore that all life must have evolved from a single common ancestor that adopted this universal genetic code billions of years ago.

Cracking the Genetic Code

Working out the nature of the genetic code is partly an exercise in logic. There are four different bases in DNA (adenine, guanine, thymine and cytosine), but twenty kinds of amino acid in protein. This means that there cannot be a one-to-one relationship between the two: there are insufficient bases or too many amino acids. Scientists, therefore, started out on the assumption that there must be multiple combinations of bases for each amino acid. If bases were "read" in pairs,

While Morse code only uses two symbols (dots and dashes), genetic code uses four bases.

The Genetic Code

there would still not be enough. (There are only 4 x 4 = 16 possible pairs of pairs: AA, AG, AT, AC, etc.) But if bases were "read" in triplets, there would be more than enough combinations: 4 x 4 x 4 = 64 possibilities: AAA, AAG, AAT, AAC, etc. It turns out that this triplet system is right. All life on Earth translates its DNA base sequence into protein by grouping particular base triplets to encode for particular amino acids. Because there are 64 possible base triplets and 20 amino acids, it means that some kinds of amino acids are encoded by more than one kind of base triplet. For instance, both DNA base triplets AAA and AAG encode for

the amino acid called phenylalanine, and any base triplet starting with AG (AGA, AGT, AGG and AGC) encodes for the amino acid serine.

The genetic code is read sequentially always in one direction along the gene, and the cell "knows" which end is the start and which end is the finish by special marking tags on the double helix. There is no "punctuation" in the form of gaps between the triplets. If you were print the base sequence of a gene, it would appear as a continuous stream of letters (A, G, T, or C) with no spaces, commas or full stops, in an order that was specific for the gene.

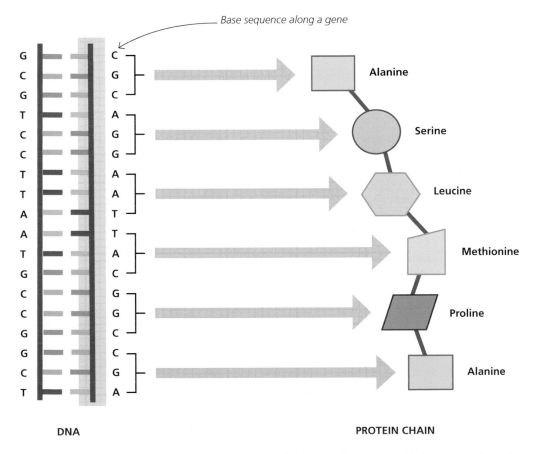

Triplets of bases in a gene encode for single amino acids in a protein. Because there are over three times as many triplet combinations as there are amino acids, some triplets encode for the same amino acid. For instance, the DNA base triplet CGC and CGA both encode for the same amino acid, alanine.

How Proteins are Made

A growing, working cell continually makes all its necessary proteins by reading and rereading its genes and using the information to link the right protein building blocks, the amino acids, in the right order.

Whenever a cell produces a protein, it needs to perform two very different processes, one after the other. First, a cell must somehow "read" its genes. Then it must use the base sequence message in those genes to build the right proteins by linking the correct amino acids in the correct order. The cell has ample microscopic machinery for doing both. The problem is that the genes and protein-making mini-machines are in two different parts of the cell.

In any cell, its DNA, and the genes it carries, are bundled together in the cell's control centre. For bacteria, the simplest kinds of organisms, this control centre is just a package of DNA in the middle of cytoplasm, but for more complex living things, including all plants and animals, the DNA is separated off in a special compartment called a nucleus. The nucleus is surrounded by membranes, just like the membrane around the entire cell. It is, in other words, like a cell within a cell. So, what about the protein-making apparatus? All cells, whether they are from bacteria or people, carry special granules called ribosomes. And these ribosomes are always in the cytoplasm surrounding the DNA, and always outside the nucleus for complex cells. Although ribosomes are so tiny they can scarcely be seen, even with the most high-powered optical microscopes, they carry everything needed to build proteins: they bring together the amino acid building blocks and the enzyme catalysts that are needed to bond them together. Any ribosome can make any kind of protein, but, critically, it needs something else: it needs the base sequence message from the genes at the cell's control centre.

Transcribing the Message

Only a tiny distance separates the cell's DNA control centre from its ribosome workshops, but that's enough to demand a special stage in the cell's message-reading hardware. Imagine that the control centre is like a library, with each gene being a book: an instruction manual for building a particular protein. (A more accurate picture would have all the books chained together, just as the genes are connected along stretches of DNA double helix.) Imagine that this is a non-lending

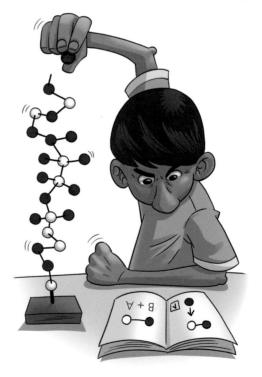

By just using four base pairs, DNA produces some incredibly complex instructions for protein building.

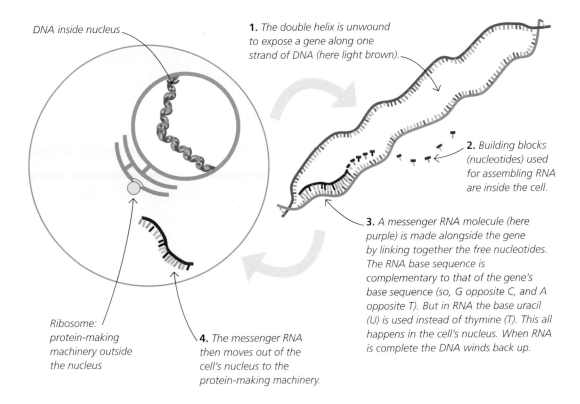

DNA inside nucleus

1. *The double helix is unwound to expose a gene along one strand of DNA (here light brown).*

2. *Building blocks (nucleotides) used for assembling RNA are inside the cell.*

3. *A messenger RNA molecule (here purple) is made alongside the gene by linking together the free nucleotides. The RNA base sequence is complementary to that of the gene's base sequence (so, G opposite C, and A opposite T). But in RNA the base uracil (U) is used instead of thymine (T). This all happens in the cell's nucleus. When RNA is complete the DNA winds back up.*

Ribosome: protein-making machinery outside the nucleus

4. *The messenger RNA then moves out of the cell's nucleus to the protein-making machinery.*

Whenever a gene is read inside the cell's nucleus, its information must first be transcribed into the base sequence along a different kind of nucleic acid called RNA. This RNA can then move out of the nucleus to the protein-making machinery inside the cell's cytoplasm.

library. You can consult each book, even take copies, but no gene-book can be removed from the library. They cannot be taken to the ribosome workshops, even that short distance away.

Copying is effectively what the cell does to solve this problem. Whenever a gene is needed, the DNA is unwound in this stretch of the double helix to expose the gene's base sequence along the critical strand. (The base sequence is equivalent to the stream of characters in the book.) Now the cell "copies" the gene. In fact, rather than an exact copy, it produces a transcript in a variant of the DNA "language": a strand of nucleic acid with a base sequence that is complementary to that on the DNA's gene. This new strand is a slightly different kind of nucleic acid, called RNA

(ribonucleic acid). But it is made in the same way as DNA – by linking together the nucleic acid's building blocks (nucleotides) already present in the cell. The nucleotides, each carrying a particular base, are joined together in a manner dictated by the base sequence of the gene. It doesn't matter that these bases are not exactly the same as the bases in the gene. If they are complementary, the specific message still gets across. For instance, a stretch of gene that reads "GCCGT" will be transcribed into "CGGCA" in the new RNA strand. RNA is built using the same kinds of bases as in DNA, except that it has U (uracil) instead of T (thymine). The RNA then moves to a ribosome, where each tiny workshop is equipped to follow the instructions in the language of RNA.

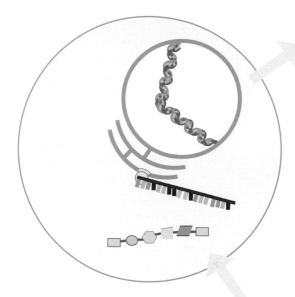

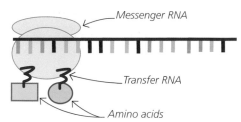

Messenger RNA

Transfer RNA

Amino acids

1. *The messenger RNA settles on a protein-making particle called a ribosome, while transfer RNA "tools" bring the first two appropriate amino acids. Here the base sequence encodes for methionine.*

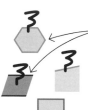

2. *Other transfer RNA "tools" get ready to collect the rest of the amino acids that will be needed to build the protein.*

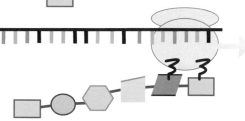

Ribosomes in the cytoplasm make up the cell's protein-making machinery. The ribosomes scan along the messenger RNA molecules, and help to bring the encoded amino acids together in the right order. Ultimately, the gene base sequence used to make a complementary RNA base sequence determines the order of amino acids, and so the kind of protein made.

3. *As the ribosome moves along the messenger RNA, more amino acids are added to the growing protein chain, according to the base sequence, until a stop sequence is met and the amino acid chain is terminated.*

Editing the Message

The RNA message made alongside the gene is not quite ready to be used to build a protein. Before it is sent to the ribosome, some editing tasks are required. Some of these involve adding more nucleotides; others take nucleotides away. First a "cap" is added to the start of the RNA message. This cap acts as a tag that can be recognized by the ribosome, so RNA and ribosome can come together to start the process of protein synthesis. Second, a "tail" is added to the end of the message. This probably helps to protect it from the ravages of enzymes on its way to the ribosome. Finally, there are "nonsense" sections of genes, called introns,

that must be removed before the instructions can be read. Introns are found in the genes of all complex cells (including those of animals and plants; they are missing from genes of bacteria). Some of these sections might have been needed for purely regulatory reasons, so that genes were only copied into RNA messages at the right places and at the right time. But their base sequences are not translatable into proteins, so they must be removed.

The Protein-Making Workshops

Each ribosome is a microscopic model of multitasking. Not only can it "read" the genetic message received on a strand of RNA, it also helps

The Genetic Code

to link amino acids together in the right order. And all this is achieved by a collective of lifeless, albeit complex, molecules.

Each RNA message could be thousands of bases long: its exact length depends upon the size of the original gene in the control centre. Once an RNA strand arrives at the ribosome workshop, the ribosome begins to read the message by clasping the RNA at one end and sliding down its length, like someone using a magnifying glass to scan the length of some small print. The base sequence must be read in the right direction to convey the correct message, and just like the original gene, a special tag directs the ribosome to start at the correct end. The genetic code always dictates that groups of three particular bases must be translated into particular amino acids – and this is the next thing that the ribosome does.

Building any protein is a demanding process. It needs precision linking of the amino acid building blocks, and the process also uses energy. Amino acids cannot stick directly to a triplet of bases.

Rather, a special transfer tool is needed to bring them together. These tools, also made of RNA, are in the cell in abundance, and a specific tool is needed for each coupling. When a ribosome is position over the message, it fits across two base triplets. This means that two transfer tools can bring down two amino acids at a time. An enzyme then bonds the amino acids together, before the ribosome moves over to the next triplet. As a result of the work of the transfer tools, the passage of the ribosome along the base sequence message coincides with the assembly of a chain of amino acids, their order dependent on the order of the bases. When the ribosome reaches the one of three particular triplets (UAA, UAG or UGA), the chain of amino acids breaks free and folds up into its special shape and the protein is ready to perform its task.

The cell is very efficient at churning out protein because more than one ribosome can work on a single RNA message at once. Sometimes a long row of ribosomes moves down the RNA, one after the other, each building the same kind of protein.

The genetic code is practically the same for all kinds of organism: the same base sequences encode for the same amino acids. Conventionally the code table lists the triplets of bases on RNA – as shown here. There are 64 possible RNA triplets – called codons – that encode for 20 possible amino acids (here given their abbreviations), with three "stop" codons telling the ribosome when a protein chain is finished.

First letter		Second letter				Third letter
		U	C	A	G	
U	UUU	Phe	UCU	Ser	UAU	Tyr	UGU	Cys	U
	UUC		UCC		UAC		UGC		C
	UUA	Leu	UCA		UAA	Stop	UGA	Stop	A
	UUG		UCG		UAG	Stop	UGG	Trp	G
C	CUU	Leu	CCU	Pro	CAU	His	CGU	Arg	U
	CUC		CCC		CAC		CGC		C
	CUA		CCA		CAA	Gin	CGA		A
	CUG		CCG		CAG		CGG		G
A	AUU	Ile	ACU	Thr	AAU	Asn	AGU	Ser	U
	AUC		ACC		AAC		AGC		C
	AUA		ACA		AAA	Lys	AGA	Arg	A
	AUG	Met	ACG		AAG		AGG		G
G	GUU	Val	GCU	Ala	GAU	Asp	GGU	Gly	U
	GUC		GCC		GAC		GGC		C
	GUA		GCA		GAA	Glu	GGA		A
	GUG		GCG		GAG		GGG		G

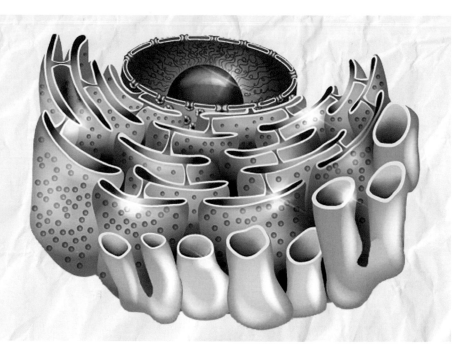

Cells that make a great deal of protein are packed with an especially large amount of protein-making machinery: countless ribosome stud the surface of a network of internal membranes called endoplasmic reticulum.

Getting the Right Proteins to the Right Place

Not surprisingly, when a body makes so many different kinds of protein, all with different functions, there needs to be a way that they all end up in the right place. Some proteins stay deep inside the cell, while others need to be attached to the surrounding membrane. Still others, such as hormones, have to be released from the cell altogether, because they need to circulate around the body in the blood.

A cell has its protein-making ribosomes loosely connected to a set of internal membranes called the endoplasmic reticulum, or ER. Particularly busy cells that have the job of making a great deal of protein are packed with lots of ER. The system can only be seen with the highest-powered electron microscopes, but protein-busy cells are packed full of it, like the layers of an onion. For the proteins made in the ER, once all the protein chains have discharged from a ribosome and folded into their final functional shapes, they are bundled together inside tiny bags of fluid, which converge on a sorting facility called

a Golgi apparatus. Here, they are finally identified and sorted for their final destination.

Enzymes needed inside the cell are sent to wherever their reactions will be required. Signalling, pumping and gateway proteins are sent to the cell membrane, while hormones and digestive enzymes are exported from the cell to perform their jobs in the fluids on the outside.

As we have seen, all the cells of the body contain the complete complement of genes needed to build that body, but genes are only active in places where their proteins are needed. Genes that are active leave their mark in a cell by a production of lots of RNA, and, so, lots of protein. Liver cells, for instance, are filled with copy after copy of the RNA message needed to make a protein called catalase, an enzyme that helps detoxify the tissues. But each liver cell, just like every other cell of the body, only contains two doses of catalase genes. The RNA and ribosome machinery can be an efficient protein-making system, even with such minimal copies of the protein instruction manual.

Chapter 5
PASSING ON GENES

Replication

DNA has a remarkable property that is crucial for life. It is able to replicate, copying its genes so that the information needed to build and maintain cells gets passed down through the generations.

Living things must make new biological materials in order to grow and reproduce. The extra material is needed when one cell divides into two, two into four, and so on. A key part of the growth of cells involves the production of more proteins, the workhorses that are encoded by the genes. But when the point is reached for cells to split in two, each new cell needs more than just some extra DNA: it needs an exact replica of all its genes. Every gene must be preserved with the same sequence of bases. For this to happen, the cell must produce an exact copy of every part of the DNA double helix. This produces a complete *bona fide* base sequence after the division for both new cells, which are called the daughter cells.

DNA replication is also at the heart of reproduction. Extra DNA is needed to produce the sex cells – the sperm and eggs – that come together at fertilization. Wherever replication happens, it must follow the usual rules of chemistry. This means that DNA needs the assistance of other molecules to complete its task: enzymes are used to bring the nucleotide building blocks together to form new chains of DNA. But this copying capability, the way DNA can produce more molecules of the same kind down to the detail of its base sequence, is otherwise unique among biological molecules.

How to Copy

The specific pairing of bases in the DNA paves the way for replication. Two strands are wrapped around each other in the famous double helix, with the bases paired between them like the rungs of a twisted ladder. The pairing is highly specific, so each sequence of bases running along one strand

is bonded to a particular sequence along the other. The A (adenine) base is only bonded to T (thymine), while G (guanine) is only bonded to C (cytosine).

Outside the living world there are other chemicals that replicate. Crystals grow bigger because their chemical units are arranged in a

When chemical crystals grow, they are replicating a shape that is determined by the specific way their atoms are bonded together. A similar trick of shape-moulding-to-shape happens during DNA replication.

DNA can produce an exact copy of itself due to the complementary two-stranded arrangement within its double helix. Each strand acts as a template for a new strand, and the entire double helix can be replicated.

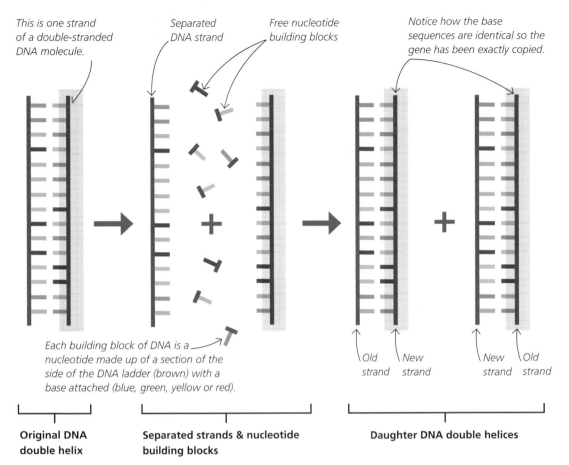

This is one strand of a double-stranded DNA molecule.

Separated DNA strand

Free nucleotide building blocks

Notice how the base sequences are identical so the gene has been exactly copied.

Each building block of DNA is a nucleotide made up of a section of the side of the DNA ladder (brown) with a base attached (blue, green, yellow or red).

Old strand   *New strand*   *New strand*   *Old strand*

Original DNA double helix

Separated strands & nucleotide building blocks

Daughter DNA double helices

highly ordered pattern, and new units added to the surface replicate this pattern: shape copying shape, copying shape. The shape of base sequence "trim" along one strand of DNA serves much the same purpose. Like the edge of a line of jigsaw pieces, it can work like a template for making a new strand. But the new strand that is made in this way is not a real copy. It has an "opposite", complementary, sequence of bases. If a single chain of DNA reproduced like a crystal, it would therefore have to go through a cycle of producing "negative" chains before going back to "positive" ones, over and over again.

However, a two-chained double helix can produce an exact copy of its entirety with each new generation of molecules. This works because both chains act as templates at the same time. Although the new chains formed are both "opposites", the complete double helices at the end have identical base pairs, with each new double helix containing one old strand and one new one. When applied to every DNA double helix molecule, this so-called semi-conservative (half-conserved) replication copies all the cell's genes, so each new cell formed after division has a complete set of genetic instructions.

Replicating the Double Helix

Before a DNA double helix can go through semi-conservative replication, it must unwind and separate its two strands, pulling apart the weak chemical bonds that exist between the bases. As always, the cell has enzymes to do this, and other kinds of enzymes are there to catalyse every subsequent stage of the entire process.

Once separated, each strand acts as a template to form a new one. Its base sequence determines how a new strand's building blocks are linked together in order, because their base pairing is specific. DNA building blocks, called nucleotides, are abundant inside the cell (just as amino acid building blocks are present so that a cell can make its proteins). A DNA strand has a "direction": each side of its "ladder" has a "front" end and a "back" end according to the way its sugar-phosphate components are arranged. (This realization gave Watson and Crick a major breakthrough in making their model of DNA in 1953.) The two old DNA strands are not only complementary in terms of their bases, but they also run in opposite directions. Like all enzymes, those involved in making DNA are very fussy: they will only make new DNA by linking nucleotides in one direction. This means that the

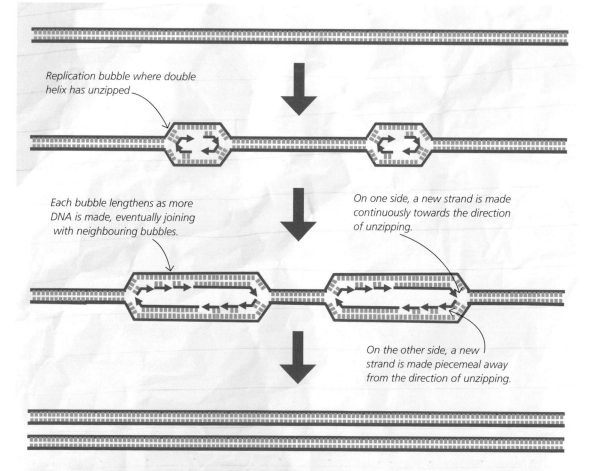

Replication bubble where double helix has unzipped

Each bubble lengthens as more DNA is made, eventually joining with neighbouring bubbles.

On one side, a new strand is made continuously towards the direction of unzipping.

On the other side, a new strand is made piecemeal away from the direction of unzipping.

Copying DNA works around numerous replication "bubbles" that lengthen and eventually coalesce to free the two new double helices.

Passing on Genes

enzymes working along each side, linking the new nucleotides, must progress in opposite directions.

There is one final twist to the story of DNA replication. During the process, a double helix does not unwind all at once. With every double helix being copied at the same time, if this were to happen, things would get very tangled up inside the cell. Instead, DNA starts its replication at several points along the chain simultaneously. This creates small "bubbles" where the strands have separated. The bubbles widen and eventually join before the two new double helices are complete. A bubble is created by unzipping the DNA on either side of a point on the double helix. Each side of the bubble looks like a fork, with one side unzipping one way and one unzipping the other way. But because the fussiness of the enzymes, only one side of a fork is ever being assembled in the direction of the unzipping. This means the opposite side needs to be assembled piecemeal, bit by bit. Each time the enzyme involved has to back-track to start again with fresh piece. Another enzyme then has the job of stitching these pieces together.

The final result of this replication is to create two double helices from one, each of which has identical base-pair sequences. And one of these sequences, of course, is made up of the precious genes that will be needed by each of the two new cells that will receive them.

Despite the size and complexity of the DNA, it is able to read and duplicate itself in little more than an hour because it is carried out in multiple places at the same time.

Speed and Proofreading

Accuracy is all-important when it comes to DNA replication. As we saw in chapter 3, if any part of a gene's base sequence is miscopied, such as by substituting one base for another, this could have catastrophic effects on the proteins subsequently made, and on the life and health of the organism. It is, therefore, all the more remarkable that DNA can replicate at such an astonishing rate, and usually without a hitch. Typically, each new strand of human DNA is built at around 50 nucleotides per second. Even at that speed, it would take a month to copy all the DNA inside a cell. In reality, it takes just an hour or so. This is because replication starts in so many places at once,

with the creation of those "bubbles". Only once all the DNA has been replicated can a cell prepare to divide: one set of replicas will go to one daughter cell, and another set to the other.

Errors of replication are so rare because the enzymes involved in linking the nucleotides to build new DNA, called DNA polymerases, have a built-in proofreading function. They can recognize when the wrong nucleotides have been inserted, and back-track to correct the mistake. However, despite these efforts, errors can and do creep into the DNA base sequence. Although many of these natural mutations are, indeed, harmful, ultimately errors of DNA replication generate the diversity of life that is at the heart of evolutionary change, as we shall see in chapters 9 and 10.

PCR is a technique that produces many copies of a sample of DNA by combining it with the appropriate building blocks and enzymes, and taking the mixture through a repeated cycle of temperature changes.

Copying DNA Artificially

In 1983, American biochemist Kary Mullis invented a way to replicate DNA in a test tube, away from the natural copying machinery that is found in every cell. The technique eventually earned him a Nobel Prize, and it provided science with a way of making trillions of copies of a specific section of DNA from tiny quantities of starting DNA. For all the ways that DNA can be studied, the bigger the sample the better. By increasing the quantity of DNA from the tiniest samples, the technique would help scientists generate more DNA for tests involved in medicine and forensic procedures.

The artificial copying method is called the polymerase chain reaction (PCR). As its name suggests, it uses DNA's natural copying enzyme, polymerase, to set off a chain reaction of self-copying. One double helix produces two, two produce four, and so on, just like the rounds of replication that take place inside a growing body. PCR uses a polymerase that comes from bacteria that live in hot springs at around 72°C (162°F). By providing the polymerase with a small starting sample of DNA and all the necessary nucleotide building blocks, PCR takes the system through a cycle of temperature changes. First, the mixture is heated to nearly boiling point, which separates the DNA's strands. Next, it is rapidly cooled to allow the first building blocks to bond to their complementary strands. Finally, it is warmed back up to the optimum hot-spring temperature that is most favourable to the polymerase. With each complete cycle, the number of DNA double helices is successively doubled.

Passing on Genes by Cloning

The multiplication of life comes down to the way cells divide, one into two, two into four, and so on. This microscopic division distributes copies of genes into new generations of cells, helping life to grow and reproduce.

As we saw in chapter 2, when cells split, their DNA coils up tighter to form thread-like bunches called chromosomes. This helps to prevent the long gene-carrying fibres from getting tangled up as the replicated DNA separates into new daughter cells. Bacteria are the only organisms whose DNA is not bundled in this way. For every other living thing, the chromosomes must perform a precise choreographic routine in the middle of the cell. This "dance of the chromosomes" is crucial to ensure that every new cell ends up with every gene.

For cells in a growing body, the dance of the chromosomes produces cells that are genetically identical. Every part of the cell's genetic makeup – the number of DNA molecules and all its genes – is preserved. The cellular process that achieves this is called mitosis. Mitosis is at the microscopic heart of anything that produces clones of organisms with complex cells, including all animals and plants. For some complex single cells, such as Amoebas, or plants that reproduce asexually through cuttings or budding, it is ultimately responsible for producing genetically identical individuals, too.

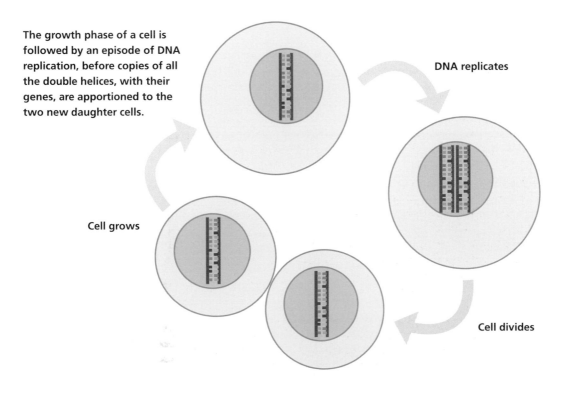

The growth phase of a cell is followed by an episode of DNA replication, before copies of all the double helices, with their genes, are apportioned to the two new daughter cells.

DNA replicates

Cell grows

Cell divides

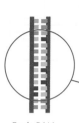

Each DNA molecule is made up of a single double helix.

1. *Before mitosis, the DNA molecules are dispersed inside the nucleus. The cell is in the growth stage between divisions, called interphase.*

Replica chromatids of a single chromosome

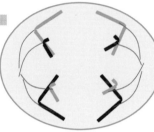

2. *Once the DNA has replicated, mitosis starts by forming chromosomes and demolishing nuclear membranes. This stage is called prophase.*

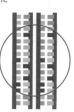

Because replication has happened, each chromosome contains two double helices in the two chromatids.

5. *A nucleus is formed around each replicated set of chromosomes. This stage is called telophase. Now mitosis is finished, the cell divides.*

4. *As the protein fibres shorten, the replicated chromosomes are split and replicas are pulled to opposite ends of the cell. This stage is called anaphase.*

3. *Protein fibres align the chromosomes along the middle of the cell. This stage is called metaphase.*

Strictly speaking, mitosis refers only to the process whereby one cellular nucleus is copied into two. This diagram shows a cell with two pairs of chromosomes. In humans, all 46 chromosomes (23 pairs) are copied and distributed in the same way.

The Cloning Dance of the Chromosomes

The appearance of chromosomes inside a cell is the first sign that it is about to divide. The triggers of the cell cycle, such as proteins called cyclins, ensure that everything happens at the right time and in the right order. By the time mitosis begins, all the cell's DNA molecules, and so all its genes, have been replicated: there are now two copies of each double helix. This means that each chromosome, in reality, materializes as two bundled threads. But these two long bundles, called sister chromatids, start off joined at somewhere along their length, and will stay like this until they are

ready to complete their separation. Depending on the chromosome, the join could be near the middle or towards the end, so each sister chromatid pair begins to look like an "X" shape, or a version of it. While all this is going on, the membranes of the cell's nucleus disintegrate so that the chromosomes can move more freely around the entire cell.

It is important that each chromosome sends one of its chromatids to one cell, and the other chromatid to the second. And the dance during mitosis goes a long way to ensuring that each chromosome does this successfully. First, every chromosome must line up along the middle of the cell, with one chromatid facing one way and its sister facing the other. Fibres made of proteins sprout from special spots on either side of the cell. The fibres connect to the chromosomes, pulling them this way and that until they are all properly aligned. By the time everything is set, these fibres span the cell with the chromosomes fixed centrally between them.

Next, the protein fibres shorten by breaking down, but their ends remain attached to the chromosomes on one side and a sprouting spot on the other. Tension builds up until something snaps: the sister chromatids split apart where they are joined and the "X" shape breaks in two. As the protein fibres continue to shrink, the newly separated chromatids are pulled to opposite sides of the cell. These are now chromosomes in their own right.

Every chromosome splits and sorts in this way, so an identical set of genes is now at opposite ends of the cell. Membranes form around them to contain the chromosomes in two new nuclei, and the chromosomes unwind back into their invisible threads of DNA. Finally, the cell pinches down the middle, and one cell becomes two. Any reasonably flexible cell divides in much the same way, but plant cells are encased in rigid cell walls, so cannot pinch in the middle. Instead, a new cell wall, with a coating of new membrane, is erected between the two new nuclei.

Asexual Reproduction

Mitosis accounts for the growth of a fertilized egg into an adult. Through successive divisions, one cell can become trillions. The copying process means that all the cells in the same body are genetically identical. But under certain circumstances, mitosis can pass replica genes through generations of bodies, too. This kind of reproduction involves making offspring without sex, and is called asexual reproduction.

For many complex single-celled organisms, such as amoebas and algae, asexual reproduction involves nothing more than ordinary mitotic cell division: cells split to form individuals that are only made up of one cell each. But for multi-celled organisms that reproduce asexually, parts of their bodies must separate. Many plants achieve this using runners or suckers: new plants sprout from horizontal underground stems or the ends of their roots. The new offshoots can mature even after the connection with the parent has withered and died. In Utah, a clonal patch of a tree called the Quaking Aspen (so-called because the wind makes its leaves rustle) covers more than 40 hectares (100 acres) and is reckoned to account for a combined weight of 6,000 tonnes of trees. Since many of these undoubtedly remain joined by their underground connections, this trembling giant is arguably one of the biggest organisms on the planet.

Cloning can be an efficient way of reproducing, but it carries risks. If every member of a population is a genetic clone, each one will also be equally susceptible to a natural disaster such as disease. The only way a colony of clones can achieve genetic variety is through gene mutation: a slow, unreliable process that more often produces harm than good. To overcome this genetic equivalent of "putting all your eggs in one basket", nature has evolved a strategy that generates variety by mixing genes into new combinations: sex.

A colony of trees known as the Pando covers a large area of forest in Utah and is made up of clones of the Quaking Aspen. At 80,000 years old, they are also, collectively, among the oldest organisms in the world.

Passing on Genes by Sex

Sexual reproduction helps to ensure that each offspring in a new generation is genetically different from its predecessors. It shuffles and mixes genes, and leads to more genetic variety in a population.

Sex involves any aspect of a biological life cycle that shuffles genes from different individuals to create new combinations. This happens most obviously whenever a sperm fertilizes an egg. The sperm and eggs have been made by different bodies with different genes, so when they join together, a third, entirely unique, collection of genes is made in next generation.

But this is only part of the story. If we go back a stage to look at how the sperm and eggs were made in the parents, we see that gene shuffling started at the very point these sex cells were being generated in the testes or ovaries. Here, a special kind of cell division ensures that every single sex cell made by a body is genetically unique. Every egg made by the woman's ovaries carries a unique combination of genes, and so does every sperm made by the man. Even though there could be 300 million sperm in each ejaculation, astonishingly, each one is genetically different from all the others. If we multiply the odds from both mother and father, the number of possible combinations of genes that could be produced at fertilization is virtually limitless. No wonder that, with the exception of identical twins, we are all so different.

This same rule of diversity applies to all other sexual organisms, covering most animals and

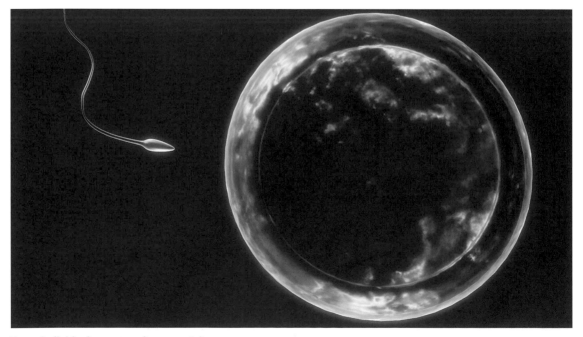

Every individual sperm and egg contains a unique combination of genes.

Meiosis always works through two successive divisions: one to separate homologous pairs of chromosomes and genes, and one to separate the chromosome's sister chromatids.

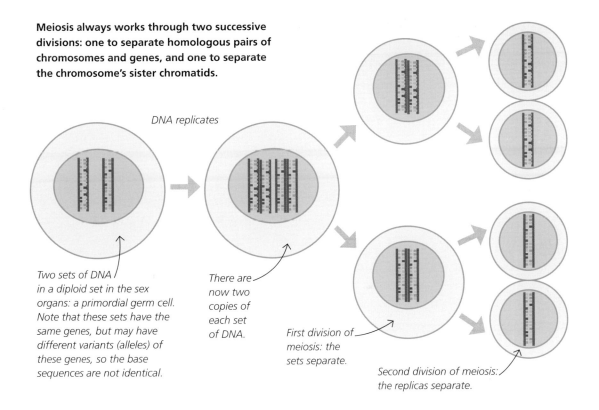

DNA replicates

Two sets of DNA in a diploid set in the sex organs: a primordial germ cell. Note that these sets have the same genes, but may have different variants (alleles) of these genes, so the base sequences are not identical.

There are now two copies of each set of DNA.

First division of meiosis: the sets separate.

Second division of meiosis: the replicas separate.

plants, and quite a few single-celled microbes too. The key to sexually generated variety, therefore, comes down not only to fertilization, but also to the special kind of cell division that produced so much diversity in the sex cells in the first place. This cell division is called meiosis and it is entirely restricted to the sex organs.

Reduction Division

Meiosis is sometimes called "reduction division" because, unlike mitosis, in which the numbers of chromosomes and genes are conserved, meiosis halves the chromosome number. If we look at the entire life cycle, this is necessary because it stops the number from doubling at each fertilization. Remember, too, that body cells contain two sets of chromosomes – the so-called diploid state. The reduction is achieved by the sets separating to form one-set, haploid sex cells. But, like mitosis, meiosis is still preceded by DNA replication to form

chromosomes with sister chromatids. This means that it must involve two divisions, one after the other. The first division reduces the diploid state to a haploid one. The second division then separates the chromatids. Note, too, that meiosis is a one-stop journey. Once the sperm or eggs are made, the only way they can pass their genes on is through fertilization. Unlike mitosis, where a cell cycle goes through potentially limitless rounds of division, one after another, mature sex cells on their own can divide no further (except in some peculiar situations where eggs of certain kinds of animals can grow without being fertilized).

Cells are only programmed to perform meiosis in sex organs. Here, special generative cells called primordial germ cells start the process. In males, the cells lie inside microscopic tubules of the testes; in females they are in the core of the ovaries. Equivalent cells in flowering plants are responsible for producing haploid pollen grains and eggs.

The Sexual Dance of the Chromosomes

The first reductive splitting of meiosis takes a longer time to happen. As well as halving the chromosome number, its special chromosome dance, exclusively performed in the context of sex, will also help produce the genetic variety that will be launched in the sperm or eggs. In order to understand how this works, we need to return to the way genes are arranged on chromosomes.

As we saw in chapter 1, the two copies of genes present in a diploid body cell may carry different variants of those genes, called alleles. In other words, if one copy, say, carries a gene for brown eyes and a gene for blood group A, the other copy could be different alleles of these same genes: perhaps for blue eyes and blood group B. The possible combinations of genes and alleles in a cell would therefore be blue–A, blue–B, brown–A or brown–B. If we add more characteristics to the mix, the number of possible combinations increases tremendously. Sexual reproduction works by jumbling up these combinations so that everyone is different. And this begins during the long first division of meiosis.

At the start of meiosis, whether it is producing sperm in testes or eggs in ovaries, the gene variants come together. They can do this because similar genes are carried on similar chromosomes, which physically pair up. These are the chromosomes that can be arranged in neat pairs when we make a karyotype from a photograph of a dividing cell. But meiosis is the only time when they actually pair up in nature. They are called homologous chromosomes and they come together perfectly at the start of meiosis. In human cells, this means that we would see 23 homologous pairs assembling from the 46 chromosomes.

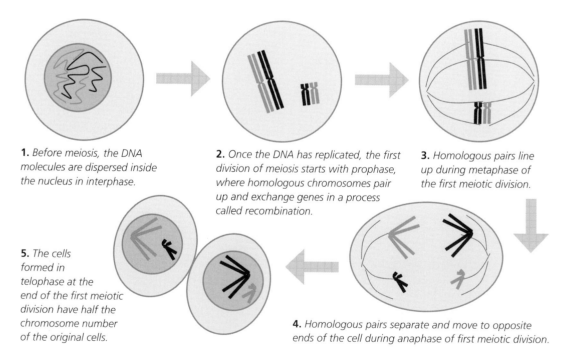

1. *Before meiosis, the DNA molecules are dispersed inside the nucleus in interphase.*

2. *Once the DNA has replicated, the first division of meiosis starts with prophase, where homologous chromosomes pair up and exchange genes in a process called recombination.*

3. *Homologous pairs line up during metaphase of the first meiotic division.*

5. *The cells formed in telophase at the end of the first meiotic division have half the chromosome number of the original cells.*

4. *Homologous pairs separate and move to opposite ends of the cell during anaphase of first meiotic division.*

By the end of the first division of meiosis, the chromosome number has halved and the cells are already genetically different. Note how there is equal chance of two "black" chromosomes ending up in the same cell, with the two "grey" chromosomes in the other.

1. *Without replicating its DNA again, the cell enters prophase of the second meiosis division.*

2. *Chromosomes align along the middle of the cell during metaphase of the second meiosis division.*

3. *The chromatids split apart during anaphase of the second meiosis division.*

4. *By the end of telophase of the second meiosis division there are four haploid sex cells.*

The second meiotic division splits apart the chromatids of the chromosomes. This means it works like mitosis, except that only half the number of chromosomes are involved.

The pairing, paradoxically, is a necessary prelude to that all-important separation that halves the chromosome number during the rest of the reduction division. The pairs come to lie in a line along the middle of the cell, with one homologue facing one way and its partner facing the other. As in mitosis, the entire dance is controlled by protein fibres. But this time, when the fibres contract, they pull the pairs apart instead of splitting individual chromosomes. As one set of chromosomes is pulled to one end of the cell, the partner set goes to the other, producing two cells with half the chromosome number that was present in the original primordial germ cell. But remember that each chromosome is still made up of sister chromatids. These are now separated in the second meiotic division to make four haploid cells in total.

The first division has actually done more than reduce the chromosome number. It has already produced genetically different cells, because one cell will receive a blue-eye allele and the other will get a brown-eye allele (located on chromosome 9).

And whether these will be accompanied by a blood group-A allele or a blood-group B (located on chromosome 15) will come down entirely to chance: it will depend on the way the homologous chromosomes were arranged along the middle of the cell before they separated. Taking all the cellular divisions into account in the testis or ovary, 50 per cent of the time a blue-eye allele will end up with blood-group A allele, so brown-eye is left with B. But 50 per cent of the time it will be the other way around: blue with B and brown with A. This means, in total, considering these two pairs of genes, a quarter of the sperm or eggs will have each of the following combinations: blue and A, blue and B, brown and A, brown and B.

Considering two gene pairs – for eye colour and blood group – gives us four possible combinations. There are two mechanisms to shuffle the alleles: genes may be on different chromosomes, and there is also recombination between genes on the same chromosome. Imagine the variety that is possible with the 20,000 genes pairs found in a human cell.

Linking and Unlinking Genes

What happens if two genes are sitting on the same chromosome? If two genes are linked in this way, it looks as though they will always get inherited together. Perhaps, say, a gene for eye colour sits on the same chromosome as a gene for hair colour. If someone has such a chromosome and it carries, specifically, an allele for blue eyes and an allele for blond hair, this surely would mean that the two characteristics would get passed down together?

It is certainly possible for genes, and so characteristics, to be linked and inherited together in this way. But there is a twist. Hundreds of genes could be linked together on the same chromosome, and this would provide a severe limit to the amount of genetic shuffling that was possible in generating that all-important variety. But cells have found a way of overcoming even this. When homologous chromosomes pair up at the start of meiosis, they can exchange segments, and so shuffle genes, even when they are linked together on the same chromosome. The process is called genetic crossing over, and it explains why meiosis can last so long. In fact, when crossing over occurs in human ovaries, the cells can remain locked in this phase for years on end, until they finally break free to produce eggs.

Crossing over is more likely to shuffle linked genes that are far apart on chromosomes. In fact, genes at the very ends of the same chromosome are shuffled so much that they behave as though they are on different chromosomes altogether. Only genes that a very close together rarely shuffle, and the ones that sit side-by-side scarcely shuffle at all (even though they are still separated by non-coding DNA: see chapter 2).

Fertilization

No-one can predict exactly which sperm will fertilize an egg: it's largely random. Humans, of course, typically produce one infant at a time. A single egg, nurtured inside one of the woman's ovaries since she was a girl (in fact, the crossing over of her chromosomes actually started when she

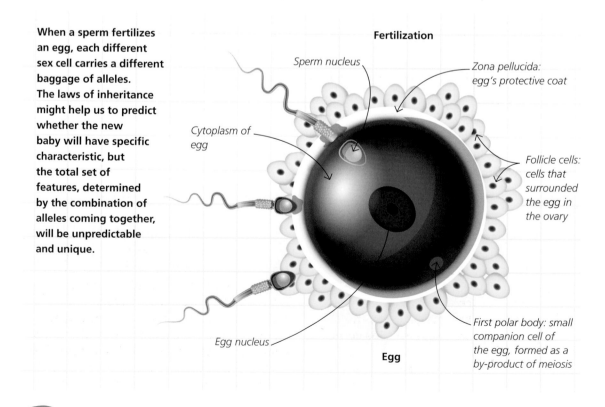

When a sperm fertilizes an egg, each different sex cell carries a different baggage of alleles. The laws of inheritance might help us to predict whether the new baby will have specific characteristic, but the total set of features, determined by the combination of alleles coming together, will be unpredictable and unique.

Fertilization

Sperm nucleus

Zona pellucida: egg's protective coat

Cytoplasm of egg

Follicle cells: cells that surrounded the egg in the ovary

Egg nucleus

First polar body: small companion cell of the egg, formed as a by-product of meiosis

Egg

What Are Identical Twins?

Human twins can arise when two eggs are released and both are fertilized in the same womb at the same time. Rarely, however, more than one person can develop from a single fertilized egg. These kinds of twins will be genetically identical, because they carry exact copies of the gene package that was present in that original cell. Each is still a 50–50 mixture of genes from either parent, but in genetic terms they are clones of each other because, essentially, they are produced by mitosis. Fertilization happens in the usual way – a single sperm fusing with a single egg – but the resulting embryo, when it is no more than a tiny ball of cells, spontaneously splits in two. Crucially, this happens at a time when all the embryo's cells are still capable of forming a complete human body, meaning that two babies develop instead of one.

was herself just a developing foetus inside her own mother's womb), is discharged into the fallopian tube, once every month after puberty. Sperm may or may not be there to fertilize it, but if they are, whichever one gets there first is the one that is used. And as the packages of genes carried by both sperm and egg are unique, so is the genetic combination in the new embryo.

Many organisms, of course, can produce many offspring at the same time. Fishes, frogs and sea anemones release hundreds or even thousands of sex cells, while flowers scatter countless pollen grains for countless waiting stigmas. But every sex cell is still genetically unique, thanks to a combination of independent assortment of gene pairs and crossing over at meiosis. Humans, rarely, can have multiple births, but twins or triplets are just as different as siblings born from different pregnancies: as long, that is, as they come from separate eggs.

Weird Things Microbes Do to Pass on their Genes

Microbes such as bacteria and viruses are too small to see with the naked eye, and are the simplest cells or particles that naturally contain genes.

In many ways, nothing matches the diversity of microbes, which are so tiny that high-power microscopes are needed just to see them. If we could reach down into their microscopic world, we would see they are as different from one another as it is possible to be. In terms of physical structure and the genes they contain, an Amoeba, a bacterium and a virus are more dissimilar than a human is from an oak tree. Many microbes, in fact, have complex cells that are similar to those of animals. Amoebas and the microscopic organisms

that cause diseases such as malaria or sleeping sickness have cells with nuclei and produce chromosomes during cell division. They can clone themselves by mitosis and even, under some circumstances, perform meiosis and the microbe equivalent of sexual fertilization. But bacteria and viruses are altogether different.

Bacteria are the smallest kinds of cell, typically about a tenth the size of the cell of an animal or a plant. Practically all of them are enclosed by a tough wall, and many have an extra slimy or oily coat on top of that. Their DNA is not bundled into a nucleus, and they do not form solid chromosomes when their cells divide. Moreover, unlike the DNA of more complex organisms, all the DNA of bacteria takes the form of closed rings.

Viruses are even smaller. In fact, they are on the very edge of life: each one is nothing more than a speck of DNA or RNA inside a protein capsule. Their genetic machinery is stripped back so much that they can only reproduce inside another organism's living cell. Without the means to feed, breathe or grow by themselves, they lack the combination of characteristics that are generally regarded as necessary to make things "alive". They are little more than encapsulated, replicating genes.

How Bacteria Have Sex

In the broadest biological terms, "sex" is defined as any process that produces new combinations of genes through shuffling and mixing. And bacteria don't even need to reproduce to perform it. Two bacteria can join together and simply swap pieces of DNA, just as if they were exchanging messages. Once they have separated, there are still only two bacterial cells, but each one contains a different set

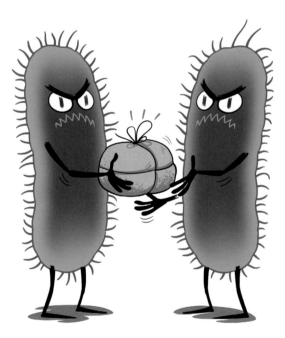

Bacteria are able to exchange genes by handing over a small package of DNA that's collected together in a ring called a plasmid.

74

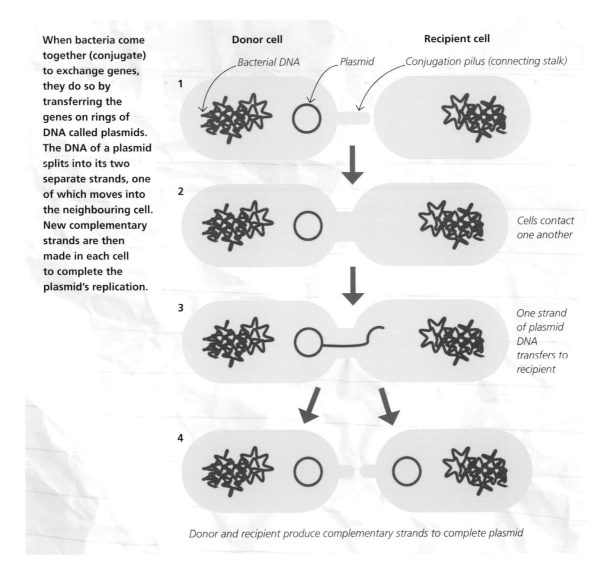

When bacteria come together (conjugate) to exchange genes, they do so by transferring the genes on rings of DNA called plasmids. The DNA of a plasmid splits into its two separate strands, one of which moves into the neighbouring cell. New complementary strands are then made in each cell to complete the plasmid's replication.

Donor cell

Recipient cell

Bacterial DNA   *Plasmid*   *Conjugation pilus (connecting stalk)*

1

2

Cells contact one another

3

One strand of plasmid DNA transfers to recipient

4

Donor and recipient produce complementary strands to complete plasmid

of genes. This sexual process, which is found only in bacteria, is called conjugation.

The critical genes of a bacterium are safely cached inside a large loop of DNA that mostly stays put. But there are other, more expendable genes on smaller rings of DNA that are scattered throughout the bacterium's cytoplasm. These rings are called plasmids. Plasmids contain genes that help bacteria to survive, and they can turn harmless bacteria into dangerous ones. Many of the genes that are carried on plasmids make the bacteria resistant to drugs such as antibiotics. Plasmids can replicate, just like the bacterium's main loop of DNA, but they can also pass from cell to cell during conjugation. They can even remain stable outside the cell for a period of time, meaning that genes can literally "leak" from one microbe to another. The rapid replication and spread of drug-busting genes is responsible for the spread of antibiotic resistance among many strains of bacteria. (This process also accounted for the experimental conversion of benign to harmful pneumonia bacteria, when Frederick Griffith was studying the chemical basis of inheritance back in 1928: see chapter 2.)

Freeloading Genes

Viruses are the ultimate freeloaders. Everything they do requires a living host cell: they cannot do anything by themselves. Virus particles can be viewed as mobile genes that move from cell to cell packaged inside a protein capsule. Viruses contain no nucleus, not even cytoplasm or membrane. And they don't even contain the usual enzymes for life. Because of this, they are generally regarded as non-living chemical particles with the potential to replicate inside living things. Many are harmless interlopers, but the effects of others are associated with some of the deadliest diseases of all, such as smallpox, rabies and ebola.

Some viruses contain DNA and others contain RNA, but no virus contains both. They all complete their replication cycle by sabotaging the cells of a living organism, and most are highly specific, targeting particular kinds of animals, plants or even bacteria. A virus's protein coat first binds to the cell membrane of its target cell, which results in it being absorbed. Once there, its goal is to use the host cell's equipment – its enzymes and ribosomes – to replicate its nucleic acid and produce more virus protein. Then the protein can encapsulate the replicas, making new virus particles, which break

out of the host cell, usually killing it in the process. The exact routine varies depending on the type of virus, and many break the fundamental genetic rules of life. For instance, so-called retroviruses carry RNA, which they use to make a DNA complementary "copy". In other words, where all cells convert DNA's message into an RNA message as a prelude for making protein, these virus rebels can also do it the other way around. Since the process is so alien to living cells, retroviruses must carry their own enzyme around with them to catalyse the process: a rare example of a *bona fide* virus enzyme. The purpose of making DNA copies is to infiltrate the host's DNA: these virus genes get stitched alongside those of the cell, so every time the cell divides, the virus's DNA is copied too.

Viruses do, indeed, appear to be alien in their nature, but the truth in their genes is more unsettling. Analysis of the base sequences has shown that they bear remarkable similarities to the cells they are infecting: viruses are apparently more closely related to their hosts than they are to each other. It seems that, in evolutionary terms, viruses are ordinary genes that have turned feral, breaking free from their hosts to become mobile, adopting their sinister freeloading cycles of replication.

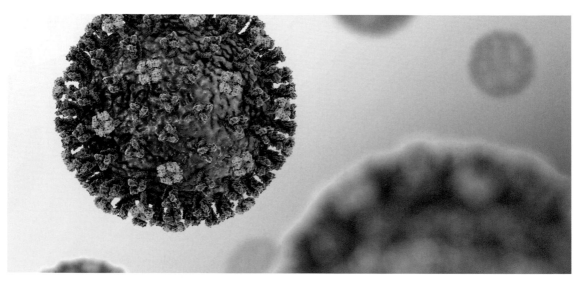

Viruses spread their genes by infecting and replicating inside living cells. Just one virus particle has to break through the body's defences for the cycle to begin.

Chapter 6
LAWS OF INHERITANCE

Inheritance Patterns

In 1866, Gregor Mendel not only revealed how inheritance worked, but also showed how patterns of inheritance could be understood in terms of the particles that we today call genes.

When Gregor Mendel wrote about breeding garden pea plants, he answered a centuries-old scientific problem: how do offspring inherit characteristics from their parents? From his results, he concluded that each kind of characteristic was determined by a particle that was passed down from one generation to another.

But Mendel went further than explaining the physical basis of heredity. By performing repeated crosses of plants that varied in ways such as flower colour, seed colour and height, he worked out that the genes were being transmitted and combined under a fixed set of rules. The rules he discovered meant he could explain the outcome of specific crosses, and how some varieties skipped a generation. These rules eventually became accepted as "Mendel's Laws of Inheritance".

Mendel's Laws are the foundation for understanding inheritance patterns today.

Mendel chose the garden pea for breeding because he could easily control its pollination.

Although some patterns of inheritance, are more complicated than Mendel supposed, his basic laws provide us with a doorway into a complex field: a starting point in trying to understand the intricate story of inheritance.

Mendel's First Law

The remarkable thing about Mendel's work is that he correctly deduced what happened with genes with little or no backup by looking at cells down a microscope. His deductions were based entirely on his interpretation of the results of his breeding experiments. His success was largely due to the careful way he took several important precautions. In the first instance, he did his best to ensure that his stocks of garden pea varieties were pure-breeding. This meant, for instance, that he started only with purple-flowered pea plants that had a credible history of only ever producing purple-flowered offspring. Second, he precisely controlled the transfer of pollen. Garden pea plants can self-pollinate, so to prevent that from happening, he secured the female flower parts under little bags. This meant that he had perfect control over the pollination process. Mendel determined the fate of the pollen, and so he also determined the fate of the crosses. Third, he replicated his crosses to ensure statistical reliability. And finally, crucially, he studied only one characteristic at a time.

Mendel began by crossing alternative varieties for one particular cross: crossed pure-breeding purple-flowered plants with pure white-flowered ones, pure tall with pure short, and so on. In each case, he discovered that only one of the varieties consistently appeared in the next generation.

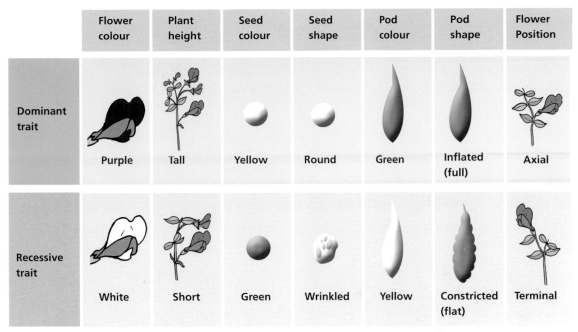

	Flower colour	Plant height	Seed colour	Seed shape	Pod colour	Pod shape	Flower Position
Dominant trait	Purple	Tall	Yellow	Round	Green	Inflated (full)	Axial
Recessive trait	White	Short	Green	Wrinkled	Yellow	Constricted (flat)	Terminal

Mendel studied seven inherited characteristics in the garden pea plant (*Pisum sativum*). Each characteristic came in two varieties. He found that the one variety (top row) was always dominant to the second, recessive, variety (bottom row).

This dominant form was clearly overriding the alternative form. He dubbed the hidden form "recessive". Then he made crosses from this first hybrid generation. The result in the second generation was that the recessive characteristic reappeared, but only in a quarter of the offspring. Mendel applied his mathematical mind to what had happened and, effectively working backwards, deduced that half of the "elements" of the recessive trait from each parent had combined in the second generation.

Based on this important bit of evidence, he came up with his first law of inheritance: the inherited "elements" (in other words, genes) from each parent came in pairs; these pairs separated into sex cells (the pollen and eggs), before combining at fertilization. The sex cells from the original parents were all genetically the same, because these plants were pure-breeding. But in the subsequent generation of hybrids, half the sex cells from each parent carried the dominant variety of the genes, and half carried the recessive variety. Today, this rule is known as "Mendel's Law of Segregation": gene (allele) pairs segregate into sex cells.

Mendel's Second Law

Mendel's first law could be articulated so clearly because his first experiments looked at one characteristic at a time. This meant that he could analyse his results in the simplest way possible to identify the patterns. But he didn't stop there. What would happen if he performed the same mathematical treatment to crosses that involved the inheritance of two characteristics at once? What would he find, for instance, if he crossed tall, round-seeded plants with short, wrinkle-seeded plants?

Again, he started with plants that were pure-breeding, but this time for both characteristics under investigation. He performed the crosses in the same way as before. He found that the results still obeyed his first law: in each case only the dominant traits (tall with round seeds) were present in the first generation, but in the second generation produced by the hybrids, he got a mixture. Looked at individually, the characteristics still obeyed the familiar three-to-one ratio: three-quarters were dominant (tall or round seeds), and a quarter recessive (short or wrinkled seeds). But when he looked at the way the characteristics had combined, he found all four possibilities in the second generation: tall with round, tall with wrinkled, short with round, and short with wrinkled. The proportions were still fixed, but this time apportioned in sixteenths rather than quarters. Nine out of sixteen had both dominant

traits (tall, round seeds), three had one dominant, one recessive (tall with wrinkled or short with round) and just one had both recessives (short with wrinkled). Mendel deduced that the same statistical principles had been carried on to these more complex crosses.

Once again, the sex cells from each original parent were genetically the same. All sex cells from the tall, round-seeded plant carried one tall gene and one round-seed gene. The other parent produced all short with wrinkled sex cells. But in the next generation, the height and seed shape

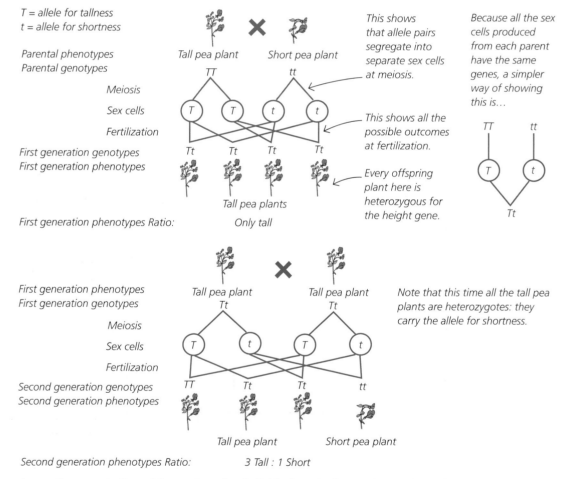

A genetic cross, starting with pure-breeding individuals and taken over two generations, can demonstrate which trait is dominant and which is recessive: the recessive one stays hidden in the first generation and has a one-in-four chance of reappearing in the second.

Laws of Inheritance

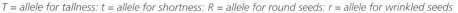

T = allele for tallness: t = allele for shortness: R = allele for round seeds: r = allele for wrinkled seeds

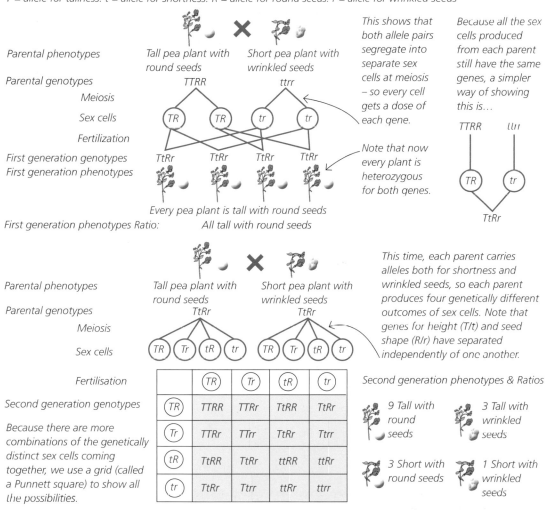

Parental phenotypes — Tall pea plant with round seeds × Short pea plant with wrinkled seeds

Parental genotypes — TTRR × ttrr

Meiosis

Sex cells — TR TR tr tr

Fertilization

First generation genotypes — TtRr TtRr TtRr TtRr
First generation phenotypes

Every pea plant is tall with round seeds

First generation phenotypes Ratio: All tall with round seeds

This shows that both allele pairs segregate into separate sex cells at meiosis – so every cell gets a dose of each gene.

Note that now every plant is heterozygous for both genes.

Because all the sex cells produced from each parent still have the same genes, a simpler way of showing this is…

TTRR ttrr
TR tr
TtRr

Parental phenotypes — Tall pea plant with round seeds × Short pea plant with wrinkled seeds

Parental genotypes — TtRr × TtRr

Meiosis

Sex cells — TR Tr tR tr   TR Tr tR tr

This time, each parent carries alleles both for shortness and wrinkled seeds, so each parent produces four genetically different outcomes of sex cells. Note that genes for height (T/t) and seed shape (R/r) have separated independently of one another.

Fertilisation

Second generation genotypes

Because there are more combinations of the genetically distinct sex cells coming together, we use a grid (called a Punnett square) to show all the possibilities.

	TR	Tr	tR	tr
TR	TTRR	TTRr	TtRR	TtRr
Tr	TTRr	TTrr	TtRr	Ttrr
tR	TtRR	TtRr	ttRR	ttRr
tr	TtRr	Ttrr	ttRr	ttrr

Second generation phenotypes & Ratios

9 Tall with round seeds

3 Tall with wrinkled seeds

3 Short with round seeds

1 Short with wrinkled seeds

A more complex genetic cross, looking at two characteristics at once, produces a more complex 9:3:3:1 ratio in the second generation. The clearest way of showing this outcome is with a grid called a Punnett square, as here.

genes separated independently of one another, so all four possible combinations were produced in equal quantities: a quarter sex cells with tall-round genes, a quarter with tall-wrinkled, a quarter short-round and a quarter short-wrinkled. When these sex cells combined to produce the generation after that, the outcome depended upon multiplying these fractions together. This, for instance, accounted for the poor show of recessive-recessive plants, because only ¼ multiplied by ¼ = 1/16 of the plants were short with wrinkled seeds. You can see each combination of characteristics explained in the diagrams shown above.

Mendel's discovery that gene pairs mingled independently like this is known as "Mendel's Law of Independent Assortment". It explains why one characteristic, such as hair colour, can be inherited independently of, say, blood group.

Genetic Crosses and Reality

Genetic cross diagrams are ways of showing the outcome of crosses that involve parents with known genotypes. We can use Mendel's Law of Segregation to predict the outcome in the offspring. It is important, however, to remember that these predictions are about the chances of getting each particular possible outcome shown, not the actual numbers produced. Many human genetic disorders, such as cystic fibrosis, are inherited recessively. This means that the only way a child can be born with the condition is if both parents have at least one dose of the faulty recessive allele. If both parents are carriers (heterozygotes) for the faulty allele, then this kind of genetic cross diagram predicts a one-in-four chance of a baby being born with the

disease. It does not mean that, if three healthy babies are conceived that the next one born will have cystic fibrosis. Rather, every baby conceived has an equal chance – 25 per cent – of having the disease.

Cross Terms and Notation

Mendel's work clarified the distinction between the genetic makeup of an organism and the actual characteristics that you can see or measure. The genetic makeup is called the genotype, while the expressed characteristic is the phenotype (such as tall and dwarf pea plants). And Mendel had shown that organisms with the same phenotypes could have different genotypes. Tall pea plants, for instance, could either be pure-breeding, or carry a single, hidden, dose of the dwarf plant allele.

The simplest way to show genotypes is by using upper- and lower-case letters for the dominant and recessive alleles respectively. Therefore, the dominant "tall" allele of the pea plant height gene might be "T", and the recessive "short" allele might be "t". There are no fixed rules on the letters that can be used, but it is wise to choose letters where the upper- and lower-case forms cannot be confused (so not M/m or V/v). And often, as here, the starting letter of the dominant characteristic is the letter of choice. Genotypes where the alleles are identical are termed "homozygotes" (or described as being "homozygous"). Genotypes where they are different are "heterozygotes" (or "heterozygous").

Mendel's Luck

As indicated above, many kinds of inheritance are too complex to be explained by Mendel's two laws alone. But Mendel's laws underpin the basic ideas that gene pairs always segregate into sex cells (so each sex cell has only one copy of the gene or allele). Remember that Mendel had no appreciation of the physical basis of the particles that he called "elements" and we call genes. This means he didn't know that many genes are tethered together on the same chromosome. If he had studied systems involving these sorts of linked genes, he would have obtained more complicated results that would doubtless have obscured the underlying laws. But he didn't. Linked genes can disobey the second law of independent assortment, because where one gene is present, its linked neighbour will follow, as long as they are not recombined when chromosomes shuffle sections by crossing over.

It was pure luck that Mendel studied garden peas. This is because all seven of the characteristics he analysed are determined by genes that reside on different garden pea chromosomes. This is all the more remarkable now we know that garden peas only have seven pairs of chromosomes in the first place. What are the chances of that happening?

Chapter 7
BEYOND MENDEL'S LAWS

Genes on Chromosomes

When Mendel's work was rediscovered after his death, scientists combined breeding experiments with studies using microscopes to show how genes were carried on chromosomes.

When Gregor Mendel died in 1884, the importance of his work had not been recognized. More than a decade later, several things happened that would bring his laws of inheritance into the limelight. At the close of the 19th century, Dutch botanist Hugo de Vries performed his own breeding experiments with plants and also concluded that inheritance depended on particles that were segregated and independently assorted when sex cells were made. De Vries only became aware of Mendel's work after he had published his own findings. At the nudging of a colleague, he acknowledged that Mendel had got there first. In 1900, English biologist William Bateson read about de Vries's experiments, including the concession to Mendel, and was immediately struck by the breakthrough. Over the next few years, Bateson became Mendel's champion and, in doing so, effectively became the major driving force for opening an unexplored avenue of science. De Vries had called his particles "pangenes" so, in 1905, Bateson coined a new word for this new branch of biology: "genetics".

It took more than 30 years for scientists to appreciate Mendel's work and his role in setting the foundations for modern genetics.

Beyond Mendel's Laws

Four figures were instrumental in the birth of twentieth-century genetics. Hugo de Vries (far left) and William Bateson (second left) oversaw the "rediscovery" of Mendel's laws of inheritance by genes, while Theodor Boveri (second right) and Walter Sutton (far right) independently devised the theory that genes are found on chromosomes.

Four years later, one of his collaborators, Wilhem Johannsen, shortened "pangene" to "gene". The gene had arrived.

By then, a flurry of scientific microscope work, and still more breeding experiments, had cast new light on the nature of these particles and their location inside cells. In particular, in an episode of remarkable coincidence, the first proof had come from researchers working on either side of the Atlantic at practically the same time.

The Chromosome Theory of Inheritance

Theodor Boveri was a German biologist who was primarily interested in studying cell biology to investigate the growth of cancerous tumours. As they still are today, sea urchins were a favourite animal in studies of cell development. These marine animals, related to starfishes, release their eggs and sperm so that fertilization can take place outside their bodies. When sperm meets egg, the resulting exposed embryos can be studied quite easily through a microscope. Earlier German scientists had discovered that cells formed chromosomes, but it was Boveri who proposed that each chromosome was effectively passed down between generations,

and that each one had its own individual identity. Through careful experiments with sea urchins, Boveri was able to show that a complete set of chromosomes was needed for their embryos to develop properly. It was enough for him to suggest that they carried the vital inherited material.

Meanwhile, in America, Walter Sutton had come to the same conclusion based on work involving grasshoppers. By dissecting cells from the sex organs of the insect and looking at them down a microscope, Sutton had discovered that chromosomes occurred in pairs, which subsequently separated during the formation of sex cells. This seemed to be a precise physical display of Mendel's First Law of Segregation.

Boveri and Sutton published the results of their work independently between 1902 and 1904, and collectively their contributions to the new field of genetics would become acknowledged as the Sutton–Boveri Theory of Chromosome Inheritance: Mendel's particles of inheritance were found on chromosomes. However, there remained pockets of resistance to the idea that chromosomes really did carry the genes. And some even doubted Mendel's laws themselves. But vindication would soon come from among these very sceptics.

The Fly Room

In 1904, Thomas Hunt Morgan was a genetics researcher working at Columbia University, USA. He was attracted to a theory devised by Hugo de Vries that evolution took place largely through a process of mutation: that new forms of life – mutations – appeared spontaneously in nature, and that these were the driving force of evolution. At the time, Morgan was unconvinced by both Mendelian inheritance and the chromosome theory.

Morgan needed an organism to study mutations. He found it in a small insect called the fruit fly. Fruit flies are tiny red-eyed insects that are attracted to over-ripe fruit; they feed and breed on the fermenting, sugary pulp and are widespread throughout the world. With the help of a team of fruit fly enthusiasts, Morgan bred the insects in their thousands inside glass bottles at a laboratory in his university that would become forever known as the "Fly Room". From around 1908, the Columbia team, led by Morgan, spent the next decade or so investigating genes and chromosomes. The work would win Morgan a Nobel Prize for establishing what he had initially doubted: that genes are carried on chromosomes.

To this day, fruit flies remain a model organism for studying genetics all around the world. But to start the process back in the early 1900s, Morgan needed to find fruit fly variants so he could study their inheritance, just as Mendel had studied the inheritance of varieties of the garden pea. It took nearly a year to get that far, but eventually Morgan's team found a mutant variation that was very clearly visible and could be traced down through the generations: the mutant flies had white eyes, rather than the usual red.

The Perfect Animals

Fruit flies (*Drosophila melanogaster*) are only about 4 mm (¹/₆ in) long, but many of their characteristics are visible to the naked eye, and they became one of the most important "model" organisms for studying genetics and other areas of biology. Males are slightly smaller than females, and notably males have a large black patch on their rear end (above). Fruit flies can easily be cultured in flasks or bottles. The material at the bottom of a flask contains the nutrients for both adults and larvae. Once the flies have laid their eggs, larvae crawl up the side of the flask to develop into pale brown, grain-like pupae before emerging as adults. Generations of flies can be sustained using this technique, which was developed by Thomas Hunt Morgan.

White-Eyed Flies

At first, the white-eyed variety of fruit fly looked as though it would obey Mendel's laws. When Morgan crossed a female red-eyed fly with a white-eyed male, the first generation of offspring all had red eyes, and the second generation after that produced three-quarters red-eyed flies with a quarter white-eyed. This all suggested that the red-eyed trait was dominant, and the white-eyed one recessive. But then Morgan found that the results depended on the sex of the original parents. If he started with a white-eyed female and red-eyed male, the results were very different: all the female offspring were red-eyed, but all the males had white eyes. The gene for eye colour was somehow linked to the sex of the flies, and the only way that Morgan could explain his results was to suggest that the gene involved was attached to special gender-determining chromosomes that had been identified a few years before: the sex chromosomes. He called this kind of inheritance sex-linkage.

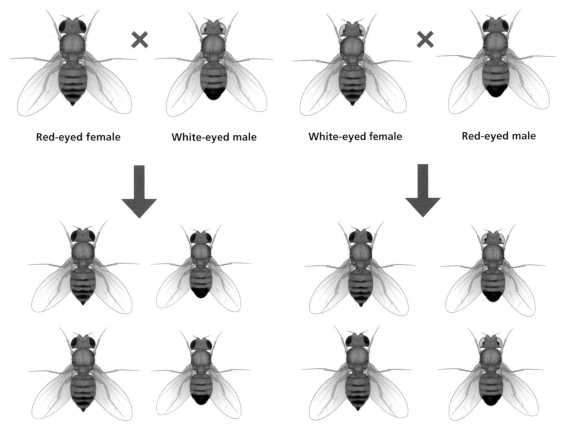

Red-eyed female × **White-eyed male**   **White-eyed female** × **Red-eyed male**

All offspring red-eyed, irrespective of sex   **All female offspring are red-eyed, all male offspring are white-eyed**

When Thomas Hunt Morgan bred fruit flies with a white-eye mutation, he found that the way it was inherited depended upon the sex of the parental flies. This was the first complication of Mendel's laws to be recognized: sex-linkage.

Explaining the Sex-Linkage Result

In crosses from red-eyed females and white-eyed males, the white-eye allele certainly seemed to be behaving recessively, since it was completely masked in the first generation of parents. But if white-eyed females were used, something about the mutation was making it persist. Perhaps females had more copies of the white-eye allele than males?

Morgan's team interpreted the results in exactly this way, and the interpretation explained the results beautifully. The gene was on a package – a chromosome – that occurred twice in females, just like all the other chromosomes in the female's body cells. But in males this particular chromosome appeared only once, so males only ever carried one copy of the white-eyed allele. The different arrangements of chromosomes clearly not only determined the sex of the insect,

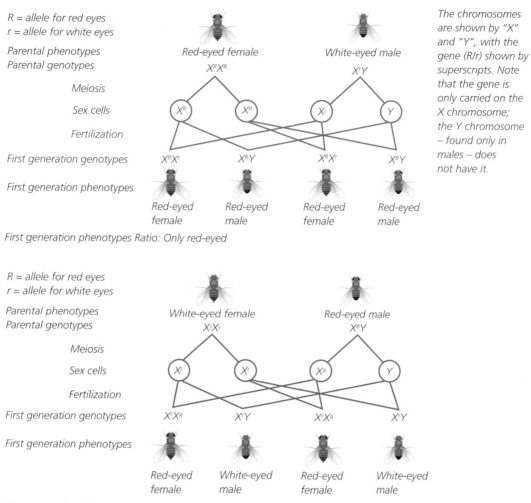

R = allele for red eyes
r = allele for white eyes

Parental phenotypes
Parental genotypes

Meiosis

Sex cells

Fertilization

First generation genotypes

First generation phenotypes

Red-eyed female
$X^R X^R$

White-eyed male
$X^r Y$

X^R X^R X^r Y

$X^R X^r$ $X^R Y$ $X^R X^r$ $X^R Y$

Red-eyed female   Red-eyed male   Red-eyed female   Red-eyed male

The chromosomes are shown by "X" and "Y", with the gene (R/r) shown by superscripts. Note that the gene is only carried on the X chromosome; the Y chromosome – found only in males – does not have it.

First generation phenotypes Ratio: Only red-eyed

R = allele for red eyes
r = allele for white eyes

Parental phenotypes
Parental genotypes

Meiosis

Sex cells

Fertilization

First generation genotypes

First generation phenotypes

White-eyed female
$X^r X^r$

Red-eyed male
$X^R Y$

X^r X^r X^R Y

$X^r X^R$ $X^r Y$ $X^r X^R$ $X^r Y$

Red-eyed female   White-eyed male   Red-eyed female   White-eyed male

First generation phenotypes Ratio: 1 Red-eyed female : 1 White-eyed male

The gene for eye colour in fruit flies is located on a sex chromosome, called the X chromosome, that occurs twice in female cells, but only once in males. It is this location that accounted for the unusual results of the breeding experiments.

but also any other characteristics, such as eye colour, whose genes were carried by it in the same way.

Through his work with fruit flies, Morgan became convinced that Mendel's idea of particulate genes and the chromosome theory were correct, and his own breeding experiments were taking genetics forwards at a fast pace. His team studied patterns of inheritance of more kinds of mutated varieties (dumpy winged, yellow-bodied, and so

on), some of which they had created artificially in the laboratory by exposing the flies to a mild dose of X-rays. Their work established that certain characteristics were inherited independently, but also that others were not. These were the ones where the genes involved were linked together on the same chromosome. Remarkably, the new genetics was not only showing that genes were firmly located on chromosomes, but it was helping to show their relative positions too.

Beyond Mendel's Laws

More About Sex Linkage

Sex-linked inheritance happens when the characteristics involved are determined by genes that sit on the sex chromosomes. As a result, the pattern of inheritance is dependent on the sex of the individuals carrying those characteristics.

Among the many people studying cells and chromosomes at the turn of the 20th century was the American geneticist Nettie Stevens. She was studying the cells of mealworms, insect grubs that metamorphosed into black beetles. Stevens had travelled to Europe to collaborate with Theodore Boveri, and he taught her the techniques he had used to stain the cells of sea urchin embryos so that you could see their chromosomes. Stevens brought these methods back to America and applied them successfully to mealworms, focusing on the differences between the sexes. She found that cells from females had 20 chromosomes, all of which matched up to form ten pairs. But in males, only nine true pairs were matched like this. The final two chromosomes were different. Stevens had discovered that (for mealworms at least) a pair of sex chromosomes determined the sex of the animal. Later, these chromosomes were dubbed XX for females and XY for males.

Following the publication of Stevens's results, chromosomal sex determination was discovered in other animals, including fruit flies and, ultimately, humans. And if one sex – the male in mealworms, fruit flies and humans – had different chromosomes, presumably these chromosomes carried different genes too. It meant that these males would only ever have one dose of sex-linked genes, something that would clearly affect the way sex-linked genes and the characteristics they determined were passed down through generations.

The three life stages of the mealworm beetle are: larva (left), pupa (middle) and imago, or adult (right).

Genes on Sex Chromosomes

Just as in fruit flies and mealworms, the sex chromosomes of humans are unmistakably different. The bigger X chromosome is just like any other chromosome in the karyotype – a long, tightly coiled bundle of DNA. But the Y chromosome is short and dumpy. In fact, it's the smallest chromosome of all. Over the years, geneticists have found that the Y chromosome, as expected, contains a kind of gene that determines maleness, but comparatively little else in terms of coding DNA. Y chromosomes, it seems, have been stripped down to the bare essentials when it comes to genetic information, carrying just 70 or so genes. The X chromosome, in comparison, is packed with many genes: around 800 in human cells. Critical genes needed for a wide range of tasks are found on the human X chromosome, including genes involved in energy-releasing metabolism, making muscle protein and allowing colour vision. This means that faulty versions of these genes are carried there too, so disorders resulting from these alleles will be sex-linked. As we saw in chapter 3, muscular dystrophy results from a non-functional allele of a gene that should otherwise instruct the manufacture of a protein called dystrophin. This protein strengthens muscles, and without it the

muscles weaken. The gene involved is on the short arm of the X chromosome. A gene on the end of the X chromosome's long arm is involved in making the visual pigments needed for us to distinguish red from green. The faulty allele, causing red–green colour-blindness, is therefore also sex-linked.

There is an important consequence of the XY arrangement in males: for these sex-linked genes, there is no back-up. As we have seen, body cells have two sets of chromosomes or genes, forming the so-called diploid state. This means that if a pair of genes contains a faulty variant, usually the normal version will make up for it. But in male

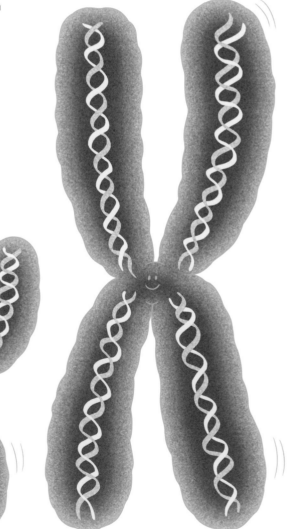

As in other organisms with chromosome-determined sex, the sex chromosomes involved are vastly different in size: the Y chromosome is much smaller than the X chromosome.

Beyond Mendel's Laws

Inheritance of Haemophilia

One particularly debilitating sex-linked disease is haemophilia. We saw in chapter 3 that this blood disorder is caused by a faulty gene encoding for protein called a blood-clotting factor. Lack of this chemical factor means that blood fails to clot properly, so even small wounds can be dangerous. The most common kinds of haemophilia are caused by genes on the X chromosome.

The normal version of the gene is dominant, while the haemophilia version is recessive. This means that girls inheriting the haemophilia variant are unlikely to be sufferers, because they are likely to have the normal variant on their second X chromosome. This will dominate and instruct the body to make the blood-clotting factor. (In fact, they produce rather less of it, but still do not have the life-threatening disease. The only way girls can have full-blown haemophilia is if they inherit two doses of the diseased gene – one from each parent – or if their second X chromosome is mutated or missing. Both situations are very rare.)

However, boys inheriting the haemophilia gene will express the disease in full because they lack the second X chromosome that should otherwise provide the dominant backup. Most famously, haemophilia blighted the European royal families from the 19th century. The haemophilia allele arose spontaneously as a mutation in Queen Victoria. We know this because there was no evidence of haemophilia in Victoria's predecessors. One of her sons, Leopold, had the full-blown disease, meaning that he inherited his single dose of the mutated gene from his mother. Because all boys inherit a Y chromosome from their father, Leopold had no "backup" allele to mask the mutation effects. He, in turn, passed it to his

daughter and grandson. Meanwhile, two of Victoria's daughters, Alice and Beatrice, were carriers for the disease. This means that, although, like Leopold, they received a single dose of the mutation, they each also received a normal-functioning "backup" from their father. However, they, too, could pass it on. As a result, each was responsible for four male haemophiliacs: in the Russian line from Alice, and in the Spanish line from Beatrice.

Queen Victoria passed on haemophilia to royal houses across Europe, including those of Spain, Russia and Germany, through two of her five daughters.

cells, genes on the X chromosome are not backed up by genes on the Y chromosome. Any recessive allele of a gene, even if it is harmful, will be expressed in males. This explains why men and boys are more likely to suffer from the effects of sex-linked genetic diseases than women and girls. The majority of the genes in cells are carried on

chromosomes that are not specifically linked to sex. These chromosomes are called autosomes and they are passed on in conventional inheritance patterns, known as autosomal inheritance. Humans have 22 pairs of chromosomes in their diploid body cells. With the two sex chromosomes they make up the total human chromosome number of 46.

More Than Two Variants

Mendel's pea plant crosses involved characteristics in which there were only two versions of each trait. But many characteristics come in more than two varieties, making inheritance less straightforward.

Even the best-known patterns of inheritance are not as simple as they first seem. Although a characteristic could be caused by a single gene at one location on a chromosome, it is important to remember that, with the exception of the XY chromosomes in males, chromosomes come in pairs. This means that genes also come in pairs, but in addition, genes come in varieties called alleles. The characteristic expressed, called the phenotype, depends on the combination of alleles, called the genotype. In his work, Mendel had documented characteristics that always came in one of two varieties: tall or dwarf, purple flower or white

flower, etc. One variety always completely dominated the other. But imagine a situation where, if two different alleles come together, the result is a third different phenotype. Now imagine that a gene isn't restricted to just two alternative versions. Think what would happen if there were three versions, or four, or more.

All these complications can and do arise in the real world of genetics, and sometimes all the complicating factors arise together. This makes patterns of inheritance far more complicated, but we can understand how by looking at one situation at a time.

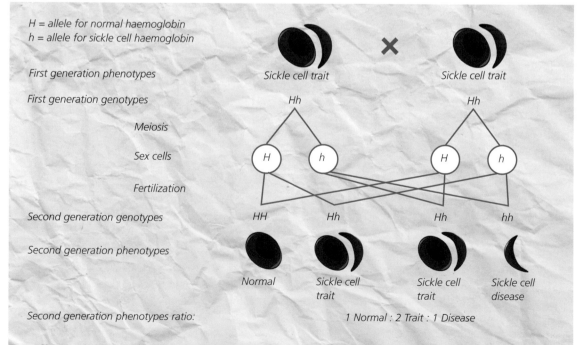

H = allele for normal haemoglobin
h = allele for sickle cell haemoglobin

First generation phenotypes — Sickle cell trait × Sickle cell trait

First generation genotypes — Hh, Hh

Meiosis

Sex cells — H, h, H, h

Fertilization

Second generation genotypes — HH, Hh, Hh, hh

Second generation phenotypes — Normal, Sickle cell trait, Sickle cell trait, Sickle cell disease

Second generation phenotypes ratio: 1 Normal : 2 Trait : 1 Disease

Sufferers of sickle cell disease can be born to parents where both mother and father have sickle cell trait.

Sickle Cell Disease and Sickle Cell Trait

In chapter 3, we saw that a gene encoding for haemoglobin has an alternative version, which instructs the formation of a variety of this protein that makes blood cells sickle-shaped. Sickle cell disease is often described as recessive, meaning that, when the two alleles occur together, the normal allele will mask the effects of the diseased one. But in fact this isn't always true. If a person carries two different alleles in their diploid body cells, even the "recessive" one can make itself known. This is because each kind of allele contributes instructions for making its special variety of haemoglobin, so cells end up with a mixture of both. As neither allele completely masks the other, the two are described as codominant. But does this cause sickling of the cells? Yes, but only under special conditions. The mixed haemoglobin only makes red blood cells sickle-shaped when the oxygen levels are low. People with mixed haemoglobin are said to have sickle cell trait, and they may only suffer from symptoms (such as cramps) when they exert themselves in exercise. But they have a distinct phenotype, and there are thus three different possible phenotypes: normal, sickle cell trait, and sickle cell disease. The inheritance is said to exhibit incomplete dominance because, in the heterozygote with both alleles, the effect of one allele does not completely mask the other.

Many traits are inherited in ways that can be described as incompletely dominant, codominant or, as with sickle cell, a combination of both.

Incomplete Dominance

In situations where traits come in clear dominant–recessive versions, one allele (the dominant version) is expressed as strongly whether it is coupled with another dominant allele or not. This is what Gregor Mendel found in his pea plant characteristics. A purple-flowering pea plant with two dominant purple-flower alleles is just as deeply purple as one coupled with a recessive white-flower allele. In modern terms, this means that the protein for

Incomplete dominance results in three distinct phenotypes. The phenotype one expressed by the heterozygotes is intermediate between the others.

making the purple pigments is produced in the same way for both. White-flowering pea plants produce no pigment-making protein at all.

Such complete dominance is not always the case. In snapdragons, one gene also determines flower colour: it makes the flower either red or white. White flowers still make no pigment. But in this case, the protein making the red pigment has half the effect in the presence of the white allele. Two red alleles together are needed to give the fully intense red flower. This kind of incomplete dominance results in a third phenotype in heterozygotes: they have pink flowers.

Although crosses involving snapdragons do not produce the same ratios as Mendel's pea plants, his principles still apply. Alleles from pure-breeding red snapdragons crossed with pure-breeding white ones will still segregate and combine from each parent to form the first generation, and their plants will still all be heterozygotes. But incomplete dominance will make all their flowers pink. If these pink hybrids are then crossed, the next generation will still produce homozygotes and heterozygotes, but a quarter of all the plants will have red flowers, half will have pink, and a quarter white.

Coat colour in rabbits is a well-studied example of a characteristic that is determined by more than two alleles.

Codominance and Blood Groups

In both dominance and incomplete dominance, the recessive allele is not expressed at all. But where codominance happens, alternative alleles, when found together in heterozygotes, are both expressed to give two different proteins at once. An everyday example of this concerns a gene that determines ABO blood groups in humans.

The ABO blood group system is based on chemical modification of a protein that is attached to the surfaces of red blood cells. Unlike the genes we discussed before, the gene encoding for this modification comes in three varieties. The A allele modifies this protein to A, the B allele modifies it to B, and the O allele makes no modification. 0 is recessive in the presence of the other two. This means that the recessive O allele is equivalent to the alleles that produce no pigment in the flowers of peas and snapdragons. But what happens when A and B come together in a heterozygote? Like the haemoglobin alleles, they are both expressed together. They are codominant, and they are expressed equally, so a heterozygous person has blood that contains an equal mixture of A and B proteins. This means that people with AA or AO alleles have blood group A, those with BB or BO have blood group B, but only AB people have blood group AB, and OO are the only ones that have blood group O.

Multiple Alleles

As well as showing codominance, the blood group system is also an example of a multiple allele system because the genes exist in more than two allele versions. Multiple alleles are responsible for highly variable characteristics. And the more alleles there are involved, the more varieties of phenotype are possible. A gene that determines coat colour in rabbits, for instance, comes in four alleles, with some dominant over others when appearing in particular combinations.

Many other human attributes are also determined by multiple variants of single genes, including those involved in pigmentation of the hair, skin and eyes. These cases are additionally complicated by the fact that such characteristics are affected by more than one kind of gene. This produces potentially a very wide variety of possible combinations, and so even more potential phenotypes in total.

Chapter 8
VARIATION

How Living Things Vary

The variety of life is rich. Not only do organisms vary across the groups, from microbes to plants and animals, but they vary within their species too. Genes play an important role in creating this variation.

Living things vary in practically every conceivable way. Across all the great kingdoms of life – animals, plants, bacteria, and others – life can be single-celled, green and leafy, or have muscles and nerves for quick motion. And a smaller, though no less significant, amount of variation exists within particular species of organisms. Walnut trees vary in height, number of leaves, and so on, just as humans vary in girth and hair colour. A lot of this variation is down to genes. Variants of genes encode for different proteins, which have different effects and lead to different characteristics. This kind of genetic variation is inherited as genes are copied, mixed up and passed down through the generations. But the environment also has a direct effect. A walnut tree that grows in shade or on poor soil could be stunted, even though it might share similar genes with one that grows on brightly lit, nutrient-enriched ground.

The contribution of genes and environment on variation is not usually an exact measure: the two interact in complex ways that not only blur the distinction, but also make studying inheritance more complicated.

Francis Galton (1822–1911) tried to apply the evolutionary theories of Charles Darwin to humans, and advocated a programme of controlled breeding whereby only elite classes should be allowed to pass on their genes.

Nature Versus Nurture

Charles Darwin's half-cousin Francis Galton was a man with many varied interests. He published on topics as diverse as weather, fingerprints and psychology, but he is, perhaps, best known for his contributions to the field of genetics, especially human genetics. Galton's book *Hereditary Genius*, published in 1869, was the culmination of years of data collection on many aspects of human biology. He measured heights and dimensions, and even devised ways to quantify intelligence and beauty (secretly ranking the beauty of the women he saw during his excursions through the country). But he also studied family relationships, concluding that intellect and attainment were at least partly genetic: eminent

Variation

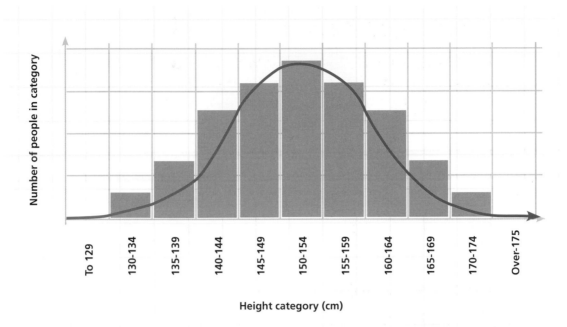

Many smoothly-varying characteristics, such as height in humans, follow a bell-shaped range, with most people having average (middling) height, and fewer having heights at the extremes of the range. This chart shows average heights of a group of women in India.

sons descended from eminent men (it was always the sons and men). Even though he acknowledged the effects of the environment, and coined the phrase "nature versus nurture" to describe the relative contribution of heredity and how people were raised, Galton was convinced that high achievement in many areas of life was passed down through the generations: character itself was determined by genes. What is more, he became equally convinced that society should take steps to improve the genetic quality of the human population. Galton coined the term "eugenics" to describe how this could be achieved, and in doing so paved the way for iniquitous 20th-century ideas about social cleansing.

Although his book failed to convince many, including Darwin, Galton persisted and went on to collect more data in an attempt to find an ancestral law of heredity. He thought he had found it in an idea that was as mathematical as Mendel's. He proposed that a precise fraction of an inherited characteristic, such as height, came from each generation of ancestors, with parents contributing half the effect, grandparents a quarter, and so on. But Galton's new law failed to account for the Mendelian ratios that were confirmed by breeding experiments that were being replicated by the turn of the century. It was becoming clear that Mendel was right all along, and Galton was wrong.

Although Galton's law of heredity turned out to be fallacious, his dogged persistence in data collection would inspire similar work in others, and his skills and techniques in statistical analysis would have a lasting impact. For instance, for continuously variable characteristics, such as human height, he confirmed that the range of variation took the form of a bell-shaped curve: most people were clustered around the middling average, with fewer and fewer towards the extremes. The idea of the bell-shaped curve was nothing new, but it was Galton's work that brought it to the forefront of genetics.

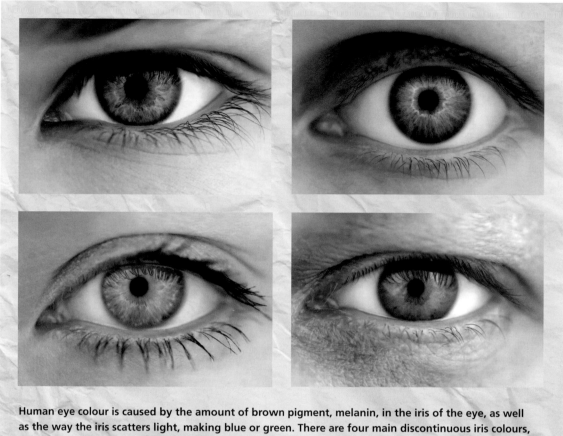

Human eye colour is caused by the amount of brown pigment, melanin, in the iris of the eye, as well as the way the iris scatters light, making blue or green. There are four main discontinuous iris colours, brown, grey, blue and green, but each one shows a degree of continuous variation in shade. Several interacting genes are responsible for these effects.

Continuous and Discontinuous

The fact that Galton and others had so much difficulty in explaining inheritance in humans is hardly surprising. At least for traits such as height and intelligence, human genetics is complicated. The simplest patterns of inheritance, which reveal any underlying laws and principles, are ones that involve discrete categories of variation, just like the pea plant characteristics that had been studied by Mendel. Mendel had seen the laws of inheritance precisely because he studied alternate versions of pea flower colour, and could calculate the simple ratios they produced in offspring from one generation to the next. This kind of variation is called discontinuous, or discrete, variation. There are just two or a few clear variants, and nothing in between. This is the kind of variation that reveals the hidden laws of inheritance the best.

Unfortunately, a great deal of variation in living things, including humans, is not discrete at all. As Galton himself demonstrated, human height comes in a range of intermediates, spread across a bell-shaped curve. The same is true of all the other characteristics that Galton tried to measure, such as intelligence, beauty or "eminence". Discrete variation in humans is limited to situations that (at least in Galton's day) would have been difficult or impossible to study, simply because their patterns had yet to be identified: things like blood groups or some genetic diseases, such as cystic fibrosis.

Variation

Factors Causing Continuous Variation

In chapter 7, we saw that single genes can come in multiple alleles, and that pairs of alleles can interact in different ways to produce many different sorts of phenotypes. With just one kind of gene, foxgloves can come in three different colours. And one blood group gene gives rise to four possible blood groups. But now imagine if a characteristic were determined by more than one kind of gene. Imagine that a characteristic is determined by genes A/a interacting with B/b, and that each possible combination, AABB, AABb, AaBB, AaBb, etc, produced a different phenotype. This might produce the sorts of intermediates that would appear to make the overall spread of variation more continuous. And the more genes that are involved, the more continuous the variation would appear. In addition, some genes have additive effects, so that for instance genes that encode for pigment could be added together to give darker colours. In reality, at least eight genes are involved in human skin pigmentation, with some determining the type of pigment being produced, and some the amount of pigment made. What is more, each gene can exist in a variety of different allelic forms, not just two. As a result, skin pigmentation varies tremendously in humans.

Environmental influences can also make characteristics more continuous. The effects of genes that help to determine height or body mass, for instance, might determine the potential of the body during development, but the precise outcome of that development will depend upon nutrition. In the same way, the sun's ultraviolet rays can increase skin pigmentation by tanning the skin, further merging any differences that might be down to genes.

By comparison, discontinuous variation, including all the variation studied by Mendel, is determined by single genes, with no additive effects and little or no environmental influence.

Like iris colour, human skin pigmentation is caused by a class of pigments called melanin. Different genes interact in a mostly additive way – simplified here – to produce a range from the darkest to the lightest skin that approaches continuous variation.

Phenotypes	Genotypes	Units of pigment
Extremely dark	AABBCC	6
Very dark	AaBBCC	5
Dark	AaBbCC	4
Intermediate	aaBbCc	3
Light	aaBbCc	2
Very light	aabbCc	1
Extremely light	aabbcc	0

It most cases, variation is caused by complex interactions between genes and is also affected by the surroundings.

The Mendelian characteristics that vary in the simplest, discrete ways are determined by single genes that come in just two allelic forms, where one allele completely dominates the other. Under these conditions, the gene determines just two variants characteristics, such as purple flowers and white flowers, or normal versus cystic fibrosis. But in many other cases, especially in human inheritance, genetics is not so straightforward. There is no single, two-allele gene that determines human skin colour, hair colour or height. And characteristics such as intelligence are not only affected by many genes at once, but also profoundly influenced by the environment. For instance, there is no such thing as "a gene for cleverness" or "a gene for musicality", although these things are undoubtedly affected by genes, .

Many characteristics once thought to be controlled by a single gene are now known to be more complex than that. Apparently trivial features, such as ear lobes and tongue-rolling, have been used as popular demonstrations of Mendel's ratios,

Producing the next Einstein or Mozart won't be just down to genetics – the person's environment will also play a key role.

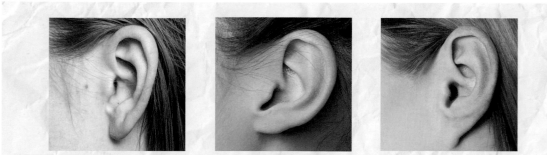

Free earlobes (left) were once thought to be a dominant trait over attached earlobes (right). Results from genetics studies, however, indicate that this trait does not follow a simple Mendelian pattern, and in reality there are many intermediate versions (middle).

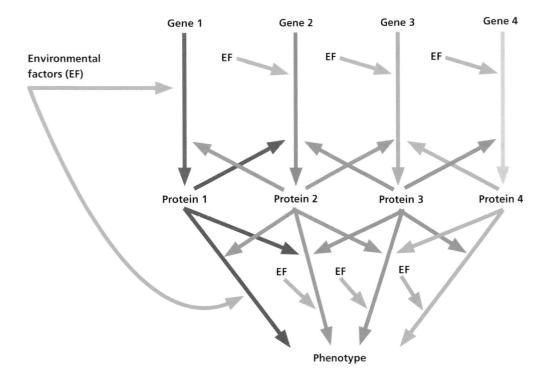

For most aspects of an organism's phenotype, many genes encode for many proteins, and these interact in complex ways to produce a particular characteristic. Some of these proteins are even involved in controlling the protein-encoding genes. On top of all that, the environment can influence the phenotype too.

but this is wrong. Free earlobes have traditionally been viewed as dominant over attached ones, but in reality there is a range of intermediates in between. Similarly, the idea that people who can roll their tongue into a tube have a dominant version of a rolling trait is wrong too. Genetics studies indicate that more than one gene is involved, so this condition is also not inherited in a simple Mendelian manner.

How Many Genes Can Determine Single Characteristics

Given the complex way that a body grows and develops, it should come as no surprise that many characteristics are, ultimately, caused by the expression of lots of genes. A single organ of the body, such as the heart or the skin, is assembled using instructions from many genes, because they rely on many different proteins that are encoded by those genes. The heart needs muscle protein, strengthening protein such as collagen, and so on. As well as collagen, the skin needs countless proteins to build and work sweat glands, hair follicles, oil glands, blood capillaries, nerve endings and pigment cells. This means that most attributes of the human skin – its hairs, its pigment, even its odour – depend on many genes.

Interacting genes are passed down in the same way as genes that have single discrete effects. But whereas single-effect genes result in simple patterns of inheritance, interacting genes rarely do. It is true that tall parents are more likely to produce tall children, but genetics can no more predict the precise proportions than it can the precise heights.

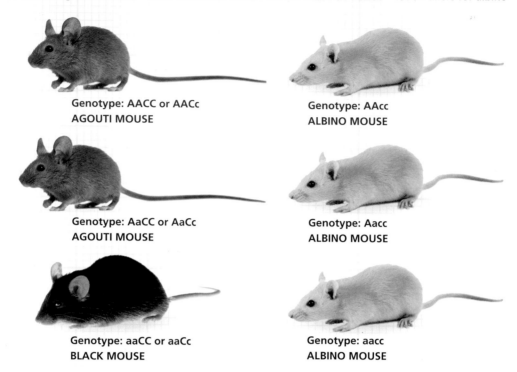

let A = allele for agouti (brown)   let a = allele for black   let C = allele for colour   let c = allele for albino

Genotype: AACC or AACc
AGOUTI MOUSE

Genotype: AAcc
ALBINO MOUSE

Genotype: AaCC or AaCc
AGOUTI MOUSE

Genotype: Aacc
ALBINO MOUSE

Genotype: aaCC or aaCc
BLACK MOUSE

Genotype: aacc
ALBINO MOUSE

Two genes are involved in coat colour in mice, and each gene comes in two allelic variants. This gives nine possible combinations, but an agouti or black mouse can only be produced if the second gene pair contains at least one dominant colour-producing allele (C).

How One Gene Can Mask Another

The fact that some allele variants of genes are dominant and others are recessive shows that genes do not work in isolation. Their expression depends on other genes inside the cell. As we have seen, dominant alleles override recessive ones by, for instance, producing sufficient functional protein to avoid ill effects. In other cases, where there is codominance or incomplete dominance, both proteins may be produced together, or the normal protein is made in insufficient quantities.

These are all examples of interactions that happen between alleles of the same kind of gene. But completely different genes at different places (loci) on the DNA can also interact, further complicating patterns of inheritance. Sometimes so-called modified genes can affect whether another gene is expressed or not: a phenomenon called epistasis. For instance, the modified gene might mask another gene, a bit like a dominant allele masking a recessive one. In chapter 1, we saw that albino coats in mice are caused by a recessive allele of a gene, and that the mouse must have two "albino" alleles for its coat to be white. But this is only part of the story, because two different kinds of gene are involved. Many mammals have grey-brown fur, but when the hairs are examined under a microscope, we see that they are banded black and yellow; the overall effect is that the coat looks brown, technically called agouti. This is the "wild type" colour of many mammals that have been domesticated, including mice and rabbits. A recessive version of the coat colour gene

Variation

leaves the entire hair uniformly black. Therefore, if we just consider this single coat-colour gene, it follows simple Mendelian inheritance, with agouti coats dominant and black ones recessive. However, a second gene found on a different part of the DNA can potentially suppress this coat-colour gene. This second gene is needed for any pigment to be produced (whether yellow or black), and it too exists in dominant and recessive forms. If a mouse has two doses of the recessive form of this gene, it stops pigment production and, irrespective of the alleles of the other gene determining the colour, the mouse ends up albino. This same epistatic system, where the pigment-making gene can suppress the pigment-type gene, is also responsible for albinism in humans.

Genes With Multiple Effects

Not only can many genes interact to produce a characteristic, but single genes can affect many characteristics at the same time. This is called pleiotropy. It happens because the protein encoded by that gene is used in many different cells to produce many different effects. One of the inherited characteristics studied by Mendel was the shape of the pea seeds: round-seed plants are dominant, wrinkly-seed plants are recessive.

As usual, the difference is related to activity of an encoded protein: wrinkled peas arise because they lack an enzyme that changes sugar into stored starch. As a result, wrinkled seeds are sweeter and contain smaller starch grains. The shape, sweetness and starch content of the peas are all pleiotropic effects of the same gene. Sometimes the association between two characteristics is less well understood, but genetics still points to a single gene being involved. For instance, 40 per cent of white-furred cats with blue eyes are also deaf. The disparate characteristics are traced to a single gene. It is thought that lack of pigment is somehow linked to a lack of fluid in the coils of the inner ear, essential for hearing. The exact protein association is unclear.

Pleiotropic effects of genes are, perhaps, most commonly revealed when malfunctioning genes have multiple effects on the entire body. Phenylketonuria is a disease in humans caused when a single gene fails to produce a critical enzyme that changes one kind of amino acid into another: phenylalanine into tyrosine. An overload of phenylalanine (and drop in tyrosine), in turn, has multiple effects, including learning disabilities and eczema. Skin pigmentation is also reduced since tyrosine is used by another enzyme for making pigment.

Some single genes have multiple effects.
A gene that makes pea seeds wrinkled also makes them sweeter and less starchy.
A gene the results in white blue-eyed cats can also make them deaf in one or both ears.

Effects of the Environment

Sometimes the environment can have a direct effect on the gene-protein system. Siamese cats have a so-called point-pattern in their colouration: the "points" of their body, their face, ears, legs and tail, are darker than the rest the body. The pattern is caused by an allele variant of the enzyme needed for making dark pigment from tyrosine; this is the same amino acid that diminishes in phenylketonuria. The Siamese point allele produces a version of the enzyme that is temperature-sensitive. It only goes to work below 33°C (91°F), which is lower than the core body temperature of a cat, but similar to the temperature at its extremities. Just like other genes, the pigment-producing gene of the Siamese cat produces pigment-making enzyme throughout its body. But the enzyme only becomes active in its coldest parts, producing dark face, ears, legs and tail.

In most cases, the environment has a much less direct effect on gene-protein systems. In fact, the effects of the environment can be subtle and difficult to interpret, mingling with the influences of genes in inscrutable ways. All living things are sustained by chemical resources, such as food and oxygen, from their surroundings. If bodies are deprived of critical requirements, their growth is stunted. Mendel studied alternative genetic varieties of pea plants – tall and short – but had to be careful to ensure that both plants were supplied with the same nutrients to reach their full potential before the inherited differences could be noted. But typically, for all plants and animals, many genes will affect growth and these, in turn, will interact with environmental factors, such as minerals and other nutrients, to produce a range of body sizes. And a whole range of other environmental influences will set to work as living things grow older. In humans, many aspects of upbringing can have a profound influence not only on growth, but health and developing behaviour.

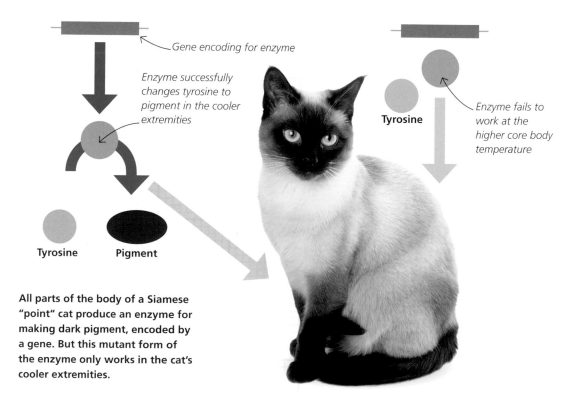

Gene encoding for enzyme

Enzyme successfully changes tyrosine to pigment in the cooler extremities

Tyrosine

Enzyme fails to work at the higher core body temperature

Tyrosine   **Pigment**

All parts of the body of a Siamese "point" cat produce an enzyme for making dark pigment, encoded by a gene. But this mutant form of the enzyme only works in the cat's cooler extremities.

Behaviour and a person's environment play an important part in influencing how a person develops and even genetically identical twins can grow up to look very different.

Heritability

Long before the time of Francis Galton, scientists had been interested in understanding the exact role of inheritance in determining the everyday characteristics of living things. Implicit in this is the question: how much variation is down to inheritance and how much is due to the environment? In other words, what really are the contributions of "nature and nurture"? Today, the fraction of the total variation for a particular characteristic that comes down to inherited genes is called heritability, and it is notoriously difficult to measure. One of the best ways of studying it is by looking at groups of individuals where the genetic variation is zero: in humans, identical twins.

Because identical twins have developed from the same fertilized egg, they are genetically identical. Twin studies could be instructive on two levels. First, if any characteristic is consistently shared between identical twins, then there is a good chance that the characteristic is genetic and not environmental: the heritability is high. Such twins will always have the same blood group, the same eye colour, the same gene-caused disorder, and so on. This all tells us that these particular traits are strongly influenced by genes. By the same token, any characteristic that persists in both as they grow older, even if they are separated and brought up under different conditions, is also likely to be genetic. Secondly, and the flipside of this, where identical twins are raised apart, any differences between them can probably be attributed to environment, and not genes. One might end up overweight, the other underweight, because of environmental influences such as diet or disease, even though they both started out with the same genes giving them the same genetic tendency to weight gain or loss.

Are Behaviour or Personality Genetic?

This chapter provides an overview of the complications involved in many characteristics that are known to be determined by genes. But what about characteristics where there might be some doubt? Stories often appear in the media about newly discovered genes that are purported to be involved in temperament, sexuality or other aspects of behaviour, and even genes for high intelligence, or genes for "criminality". Can there really be genes for such complex characteristics?

Every aspect of behaviour, more or less, is affected by genes. Behaviour emerges as a product of our nervous system, which itself is a product of a complex process of development that is ultimately controlled by genes. But, as we have seen throughout this chapter, there is rarely a simple relationship between a single gene and a single, well-defined attribute. Instead, genes interact with other genes to be expressed in different ways, and many will only show their effects under particular environmental conditions. As we will see in chapter 11, some variants of

genes have been identified that appear to predominate in particular groups of people, such as those with high IQ or psychopaths. But this does not mean that these genes will always govern behaviour in a way that is predictable.

Penetrance

Some genes always end up determining particular characteristics. Pea plants that carry two copies of the white-flower allele will always produce white flowers, while mice with the appropriate combinations of genes will end up with brown, black or albino coats. In the same way, there is 100 per cent chance of someone with the cystic fibrosis genotype going on to develop the symptoms of the disease. But not all genes have such a predictable outcome.

As its name suggests, a tumour-suppressor gene called BRCA1, stops tumour growth. It encodes for a protein that repairs damaged DNA, and can even help destroy cells that are irreparable. In doing this, it prevents cells from breaking the control of the cell cycle, stoppings cancerous growth. The mutated allele of this gene fails to do this, and increases the risk of breast cancer. But, compared

with the cystic fibrosis allele, its effects are not so predictable. Overall, women with the mutation have an 80 per cent chance of developing the disease during their life, but whether or not this will happen – and when – seems to be influenced by factors such as lifestyle and diet. Sometimes the onset of genetically-related diseases even just looks like a matter of chance and bad luck.

The proportion of individuals in a population that show the effects or symptoms of a gene is called the "penetrance" of the gene. This means that, across a lifetime, the BRCA1 gene has a penetrance of 80 per cent, and cystic fibrosis 100 per cent. The penetrance of many genetic disorders is age-related, with the risk of developing the symptoms rising with time. Huntington's disease, for instance, a degenerative disease of the brain, develops only in maturity, but its penetrance rises to 100 per cent by the age of 70.

Epigenetics

The most striking way that the environment can affect characteristics is to affect genes directly. Studies over the last two decades suggest that this is a normal part of the development process.

Like any chemical in the body, DNA is not impervious to change. It can react just like other chemicals, and not only in the normal ways designed to replicate genes and make proteins. Some of these reactive influences can be very damaging. As we will see in more detail in the next chapter, outside influences can change DNA by interfering with replication. These changes, called mutations, even occasionally happen by spontaneous error. Thousands of trillions of DNA building blocks (the nucleotides and bases) are joined together in a typical body at any one time, so it is hardly surprising that the occasional wrong unit is installed. When such errors are incorporated into the base sequence of DNA, these same errors could be replicated whenever the new error-ridden DNA is copied. And that means, the altered genes could get inherited.

But is there any way such interferences can affect the genes without altering the base sequence? Traditional dogma says no. It would be like saying that potentially inherited characteristics could be acquired from the environment, while keeping the genes the same. But, remarkably, studies suggest that such a thing is possible. These effects are called epigenetic (literally meaning "outside genetics") changes. They involve chemicals that do not change the base sequence – the instructions – of a gene, but rather change the way the gene is read, or expressed.

Epigenetic effects mean that the same genes are still present in the DNA, but they have been chemically modified, so they are less easily read.

The Effects of Methyl Groups

The tiniest possible fragment of organic matter is a methyl group. Like a molecule of the natural gas methane, a methyl group contains a single carbon atom. In methane, the carbon is surrounded by four hydrogen atoms, but a methyl group has just three. The missing hydrogen makes it reactive: desperate to cling to another piece of organic matter. Methyl groups are common in cells, and are involved in transferring carbon atoms from one molecule to another. And DNA is a favourite target.

Remember that the DNA double helix contains a sequence of base pairs that are the "rungs" of the twisted DNA "ladder". Remember, too, that there are four kinds of bases, and that the sequence of bases along one side carries the genetic instructions of the genes. One of these bases, called cytosine, is always bonded with a base called guanine, and this is the specific target of methyl groups. Notably, they usually bond to cytosine, and always at one particular place. This could be viewed as an unfortunate accident, like the effect of a mutation. But the addition of the methyl group, called methylation, is actually catalysed by an enzyme made by the cell, suggesting that it is very deliberate.

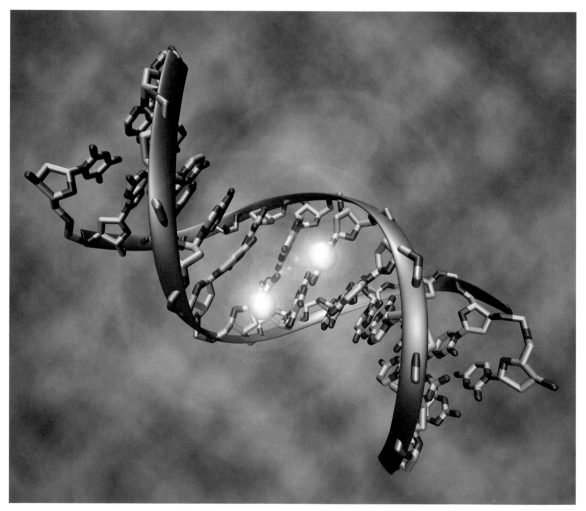

This image shows a DNA molecule that is methylated on both strands on the centre cytosine. DNA methylation plays an important role in epigenetic gene regulation, development and cancer.

Variation

Methylation is a chemical reaction that affects DNA and DNA-packaging proteins to deactivate genes. Recent research suggests that this kind of epigenetic effect could be important in helping to control aspects of development.

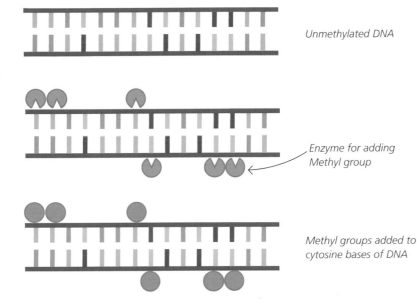

Unmethylated DNA

Enzyme for adding Methyl group

Methyl groups added to cytosine bases of DNA

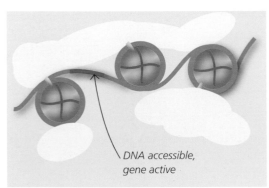

DNA accessible, gene active

Without methyl groups added to DNA's packaging (histone) proteins, gene is active

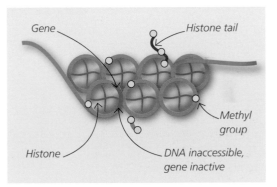

Gene

Histone tail

Methyl group

Histone

DNA inaccessible, gene inactive

With methyl groups added to DNA's packaging (histone) proteins, gene is inactive

The effect of DNA methylation is to switch off the gene. Specifically, the methyl groups deactivate parts of the DNA that initiate the process of producing an RNA "copy" needed for making the protein. But, overall, the effect is dramatic: it silences the gene. Methylation is now known to be a widespread feature of other aspects of the cell's protein-making system, including the RNA "copies" and proteins involved in packaging the double helix. These proteins, called histones, are responsible for coiling the DNA tighter to make

chromosomes. But they are also involved in loosening the DNA so that its genes can be read. If they are methylated, the DNA isn't loosened, so genes are silenced.

The fact that cells methylate their genetic systems in this way suggests that it serves an important purpose in regulating the activity of genes. Genes must be switched on or off as a natural part of an organism's development. This ensures that genes are only activated in the cells where their encoded proteins are needed.

Some experiments have shown that diet can have epigenetic effects in mice, with genetic changes being passed down from parents to offspring over a few generations.

Can Epigenetic Effects Be Inherited?

Theoretically, it is possible for epigenetic effects to be passed down through the generations. A chemical tag such as a methyl group, stuck to DNA to silence a gene, might be transmitted that way through sperm or eggs to an embryo. But does this really happen? Some experiments suggest that it can.

Mice, for instance, appear to have their genes switched epigenetically through diet alone, with an effect that persists from parents to offspring. But so far, these epigenetic effects seem not to be long-lasting: they peter out within a few generations. And, remember, that these are only subtle alterations that involve switching genes on or off. The base sequence, which carries the instructions of a gene, stays exactly the same.

Chapter 9
MUTATION

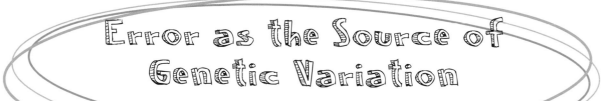

Error as the Source of Genetic Variation

When cells copy their genes and go through cycles of division, occasional mistakes creep in. These mutations create life's extraordinary diversity.

Cells have the potential to divide and reproduce at a phenomenal rate. Splitting once every 20 minutes, bacteria can produce millions of cells from one in just 24 hours. Animal and plant cells, including our own, typically divide more slowly, but even so, it is easy to think of them as microscopic replication machines. As we have seen, systems are in place to help ensure that everything goes to plan. This means that genes are copied accurately and, when it is time for a cell to divide, the chromosomes are sorted properly. During growth or asexual reproduction, genes and chromosome number are conserved down generations of cells. When sex cells are made, genes are shuffled around and the chromosome number is halved, but the next generation will still end up with a complete set of all genes.

But errors can happen. DNA might be miscopied or chromosomes mis-sorted. Cells can be produced with genes that carry a slightly different base sequence, while chromosomes, or sometimes just bits of chromosomes, can end up in the wrong place. When this happens, cells are left with too many of some genes, or a deficit of others. All these mistakes are called mutations.

When Can Mutation Happen?

Two aspects of a cell's life can go wrong in a way that could affect genes. First, DNA could be miscopied during its replication: the nucleotide building blocks linked in the wrong order, altering the DNA's base sequence. This is called a gene mutation. Second, there could be an error in the choreography of the chromosomal dance associated with cell division, during either mitosis or meiosis. Perhaps a chromosome moves into the wrong position. As a result, daughter cells do not

Mutations are very rare: the chance of any base pair in a human gene being miscopied across an entire year is reckoned to be one in half a billion. That doesn't sound much, but when you consider the trillions of cells that make up a human body, and how many divisions are involved, mutations will inevitably happen somewhere.

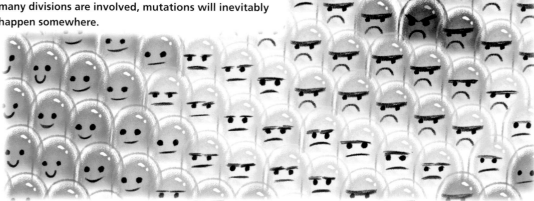

Mutation

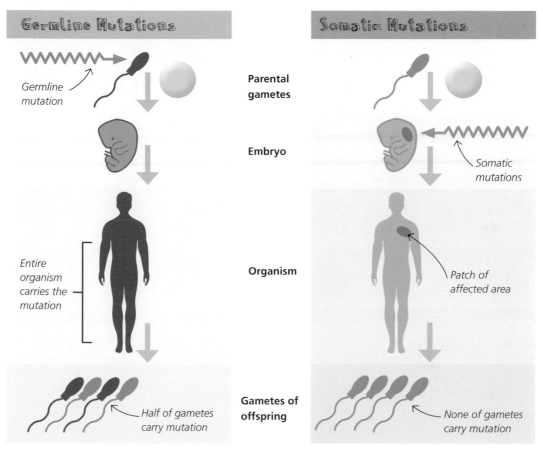

Germline Mutations

Germline mutation

Parental gametes

Embryo

Entire organism carries the mutation

Organism

Half of gametes carry mutation

Gametes of offspring

Somatic Mutations

Somatic mutations

Patch of affected area

None of gametes carry mutation

Genes, and so mutations, that are passed on from parents to offspring are said to be in the germline: "germ" cells being an old term for sex cells. When errors only affect other parts of the body, they are called somatic mutations.

receive their full complement of chromosomes. This is called a chromosome mutation. Chromosome mutations still affect genes because entire chromosomes, carrying hundreds or thousands of genes, end up in the wrong place.

Many mutations are recognized by the body and dealt with straight away. As we have seen, the cells' molecular workhorses, enzymes that are involved in replication, can spot the error and backtrack to correct it. And cells that are deemed to be just too defective can be sacrificed before the mutation spreads. But sometimes mutations can sneak through the body's natural screening process.

Mutations that persist when a body grows will not affect the next generation of offspring unless they involve the organs that make sex cells, such as sperm or eggs. Otherwise, a mutation will only affect the part of the body in which the mutated cells are dividing. The effect is that we end up with a body that is a genetic mosaic: it is made up of patches of genetically different cells, with each individual patch arising from a single mutated cell. If mutations happen when sex cells are made, they can then be passed on to offspring. This is the so-called germline of cells and genes. And only mutations in the germline can be inherited from generation to generation.

What We Lose With Age

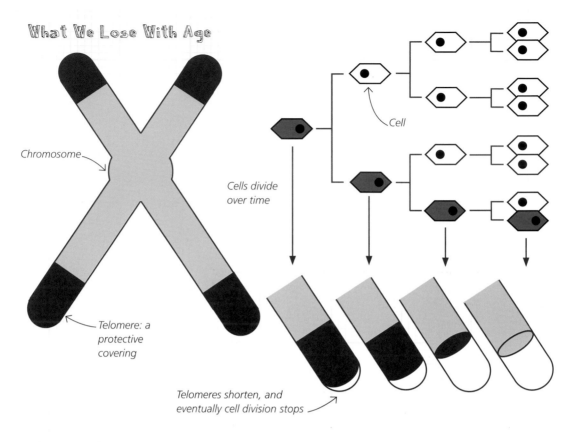

Chromosome

Telomere: a protective covering

Cell

Cells divide over time

Telomeres shorten, and eventually cell division stops

In humans and other backboned animals, the telomeres on chromosomes consist of short base sequences repeated hundreds to thousands of times. They have shrunk to around a third of their original length by old age.

Mutation and Ageing

Normal cells in the body are not immortal. Even though there are many systems in place to repair mistakes and damage, inevitably cell performance declines as the body gets older. Typically, in a multi-celled body, this happens after about 50 rounds of division. Much of this is probably associated with accumulated mutations that affect the workings of the cell. Proteins fold slightly differently, so cannot perform their tasks properly, while energy-releasing systems and nutrient uptake become less efficient.

However, one aspect of chromosome behaviour has a particular link with ageing. The tips of

chromosomes are "capped" with special non-coding pieces of DNA called telomeres. These caps help to protect the vital coding parts of the DNA that lie deeper inside each chromosome. But, as cells progress through cycle after cycle of division, the caps erode, and eventually a point is reached whereby the underlying coding regions could also become degraded. As a result of this, cell performance declines more steeply.

Telomeres help to explain why normal cells and organisms cannot last for ever. But the genes they carry are potentially immortal. Genes are copied and transmitted from parents to offspring and, in their replica forms, can last for many generations.

Creating Variety

Practically everything about all the different life forms on this planet points to them having descended from a single common ancestor. Without exception, all known cells, from bacteria to animals and plants, use the same kinds of molecular tools to breed and keep themselves alive. They use DNA and RNA to carry their inherited information, use proteins to perform tasks, and even use the same genetic code to make one from the other. This deep similarity goes far beyond coincidence: it indicates that everything has evolved from a single point. That common ancestor was undoubtedly a single cell, and we will explore some theories about what it may have been like in chapter 10. But, for now, it is enough to appreciate that life began with simplicity and just a few genes.

Today, even the simplest bacterium contains thousands of genes, and more complex, multi-celled life forms have tens of thousands. So where did all these genes come from? The answer is mutation. Although most DNA copying errors lead to something malfunctioning, if enough of them take place over enough time, some will have had beneficial effects. Genes found in modern organisms that were not found in earlier life, such as those involved in vision, or thought, or keeping cells together in a multi-cellular body, are the descendants of more ancient ones found in the very first cells of life. Over billions of years, and across countless rounds of cell division, errors that persisted produced new kinds of genes that, in turn, mutated still further.

During a period known as the Cambrian Explosion about 540 million years ago, many new organisms appeared, including this trilobyte. Mutation provided the new genes to produce this diversity of life forms.

Gene Mutations

Information in genes takes the form of a sequence of building-block units, called bases. If the sequence of bases is altered, the resulting gene mutation affects the organism in ways that vary from the trivial to the catastrophic.

You will remember that, at the level of DNA molecules, a gene is a length of the double helix. More specifically, it is a stretch of one side of the twisted "ladder", in which there is a particular sequence of chemical units called bases. The bases make up the "rungs" of the ladder, and a gene consists of the bases down one side of the rung sequence. Bases come in four types, usually denoted by their letter abbreviations: A (adenine), T (thymine), G (guanine) and C (cytosine). The

During the replication stage, the DNA can mutate, with base pairs being added, lost or replaced completely.

sequence of bases is critical because it determines the kind of protein made by the cell. And proteins are the molecular workhorses, doing all the jobs of life, from driving crucial chemical reactions to moving substances from place to place. If the gene's base sequence is altered, this changes the sequence of protein building blocks (called amino acids), so the protein fails to assume the shape necessary for its task. A gene can be tens of thousands of bases long, but often just one base change is needed to disrupt the protein in this way. As we saw in chapter 3, it can result in a genetic disease, such as sickle cell or cystic fibrosis.

Gene mutations can happen when DNA is replicated. This involves the two sides of the DNA double helix separating, with new sides assembled alongside the old ones by lining together DNA building blocks. Each of these building blocks, called nucleotides, carries a specific base, and they are linked in a specific order dictated by the exposed bases on the old strand of the double helix. See chapter 5 for more about this. But if the wrong nucleotides (and, therefore, wrong bases) are installed, the result is a gene mutation.

Kinds of Gene Mutation

As far as the entire mechanism is concerned, the most straightforward kind of gene mutation substitutes one base pair for another. This single-point change to the gene's base sequence could end up substituting an amino acid in the protein. This is because each different kind of amino acid in the protein is encoded by a triplet of bases in the gene. For instance, the triplet "CCC" in a gene encodes for the amino acid glycine, but if it mutates to "TCC", it encodes for arginine instead.

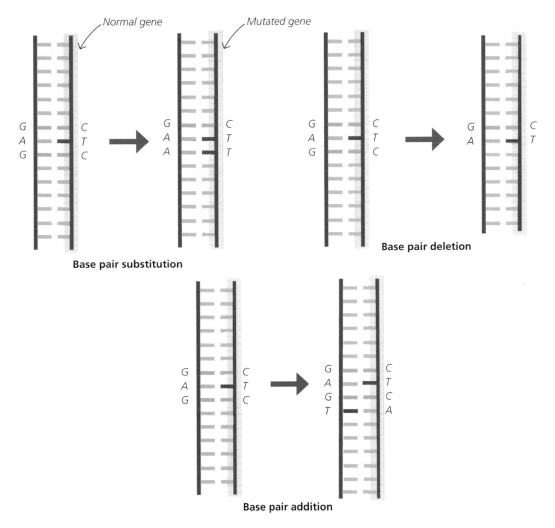

Gene mutations involve alterations in the DNA base sequence by substituting bases, deleting bases, or adding them.

Even in a protein that is a hundred amino acids long, this single change might make the protein chain fold into a different shape, so it would be unable to properly perform its assigned task.

Mutations that involve adding or deleting bases are usually much more devastating. These are typically frameshift mutations. If a single base is added (called an insertion) or removed (called a deletion), this upsets the entire reading frame after that point. This is because base triplets are read sequentially. Imagine, for instance, that we started with the following base sequence (subdivided into triplets so you can see the pattern) "AAA-AAC-AAT-CGC", and added "C" to the first triplet. The cell would then read the sequence as "AAC-AAA-CAA-TCG-C". In other words, all the triplets after the insertion would be shifted, affecting more amino acids in the protein. Frameshift mutations may result in proteins not being made at all.

What Causes Gene Mutations?

Most gene mutations happen spontaneously. They occur when DNA is replicated or when it is being repaired. At the level of the molecules, a number of chemical effects can introduce random changes to a gene's base sequence. Sometimes these start with something as apparently innocuous as a wrongly positioned hydrogen atom. Like all organic molecules, DNA is coated in hydrogen atoms, and even a single misplaced hydrogen atom in a "rogue" base can make the bases pair up wrongly. In other cases, the bases have been altered in a more dramatic way, or the two strands of the double helix "slip" position with respect to one another. In all these situations, a mutation results from the simple fact that a tiny fraction of the chemical reactions involved will not be perfect.

But sometimes it is something in the surroundings that will tip the balance. These environmental influences are called mutagenic effects and come in two forms: chemical or radiation. Plenty of chemicals actually react with components of DNA's double helix, thereby upsetting its structure. Some of these are molecules that resemble DNA's bases in physical structure. This means they can be mistaken for genuine bases by the cell (more specifically, by the cell's catalysing enzymes) and end up being used as building blocks instead of the real ones. But their differences from the real thing mean that replication does not happen properly and the base sequence is altered.

The energy from radiation can knock the various components of DNA out of kilter. Gamma rays and X-rays are so powerful that they can strip electrons from atoms. Electrons are negatively charged, so the effect leaves behind positively charged patches, which can wreak havoc on the structure of a molecule. If this happens in DNA, the double helix can be so damaged, and the base sequence changed so much, that it kills the cell. Ultra-violet light is less powerful, but can still be harmful. UV light causes thymine base pairs to bond to one another, linking them together on the same chain, and so breaking the usual base-pair "rungs" of the DNA ladder. Changes such as these can buckle the DNA, stopping replication completely.

The Effects of Gene Mutations

Changes to the base sequence of a gene can have very different consequences for the organism carrying the mutation. Some mutations have no effect on wellbeing at all. But others can be so devastating that they kill the cells carrying them, or cause lasting harm to the entire body. It all depends on the effects on the proteins encoded.

The base sequence of a gene determines the sequence of amino acid building blocks in its specific encoded protein. This, in turn, determines the way the protein chain folds up to form a specific shape, as well as its chemical properties, which affects how these proteins work. As we saw in chapter 3, for the protein haemoglobin, a single base substitution that leads to the amino acid glutamic acid being swapped for valine results in sickle cell disease. However, it is possible for some substitutions to have no effect at all. "CCC" in DNA encodes for glycine, but if it mutated into "CCT", it would still encode for glycine and the protein would stay the same. This is because some base triplets encode for the same amino acid. Such mutations are said to be "silent": they have no observable effect.

Frameshift mutations (insertions or deletions) are the most damaging of all because they affect all the amino acids after the mutation point. Sometimes so much of the protein is affected that it fails to work at all, or it may not even be produced. A number of different mutations cause cystic fibrosis, some of which involve this kind of frameshift mutation. In one mutation, however, an entire triplet of bases is deleted, which results in a loss of single amino acid with no frameshift after that point.

Intuitively, any gene mutation is more likely to cause harm than good. This is because a random change to something as complex as a protein is more likely to be defective than productive; it would be like making a random change to the workings of a car or a computer. But given enough opportunity, a few mutations may be beneficial in the long term. Sometimes, by chance, they will end up producing a protein that does its job more efficiently, or even performs a different job altogether. Such persistent mutations provide the raw material for the evolution of life, as we will see in chapter 10.

As well as being inherited, gene mutations are also acquired during a person's lifetime. Sometimes this happens during cell division. Other times, DNA is damaged by environmental factors, including radiation, certain chemicals and viruses.

Radiation

UV Radiation *Both natural sunlight and tanning beds*

X-Rays *Medical, dental, airport security screening*

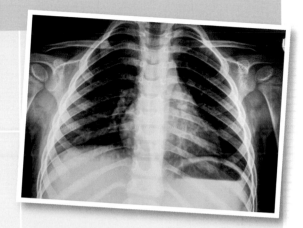

Chemicals

Cigarette smoke *Contains dozens of mutagenic chemicals*

Nitrate & nitrate preservatives *In hot dogs and other processed meals*

Barbecuing *Creates mutagenic chemicals in foods*

Benzoyl peroxide *Common ingredient in some products*

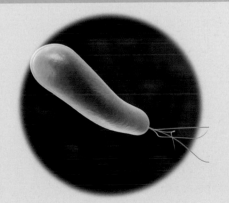

Infectious agents

Helicobacter pylori *Bacteria spread through contaminated food*

Human Papillomavirus (HPV) *Sexually transmitted virus*

Chromosome Mutations

Mistakes that happen during cell division can upset the chromosome number or result in misplaced pieces of chromosome. These chromosome mutations are visible under a microscope and affect the numbers or arrangements of genes.

The genes of any kind of organism with complex cells (including animals and plants) are precisely arranged along molecules of DNA that bundle more tightly during cell division: the bundles appear as solid threads called chromosomes. Because chromosomes appear after DNA replication, this means that each bundle contains a package of two DNA replicas. A controlled movement – a chromosome "dance" – must then take place during cell division to ensure that chromosomes are distributed properly into daughter cells. In mitosis, this involves separating the replicas to create two genetically identical cells. In meiosis, two successive divisions first separate the homologous chromosomes, then separate the replicas. Meiosis also shuffles alleles between chromosomes to mix up the gene combinations, helping to ensure that sex cells carry as much variety in the arrangement of genes as possible.

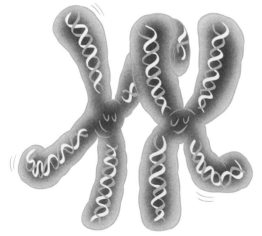

During cell division, chromosome appear to go through a complex dance routine to make sure that each daughter cell gets the right genetic information.

Chromosome mutations arise because these "dances of the chromosomes" are not always perfect. This means that daughter cells might end up with an abnormal number of chromosomes. If this happens during mitosis associated with growth of the body, the result is that some cells in the body will carry the abnormality, but others will not. Some such abnormalities prevent any further cell division, but others may proliferate. However, as with gene mutations, only mutations that happen during the formation of sex cells, during meiosis, will be passed on to offspring. Like gene mutations, some chromosome mutations can be sources of genetic variation that help to drive evolution. But many others can cause problems with development.

Polyploidy

A common type of chromosome mutation disrupts the separation of chromosomes just before daughter cells are made. During meiosis to form sex cells, chromosome number is halved: chromosome pairs line up along the middle of the cell, then separate so the partners from each pair move in opposite directions to different cells. These partners, called homologous chromosomes, carry the same kinds of genes (although the allele forms of these genes may be different), so each sperm, egg or pollen grain ends up with a single dose. In other words, meiosis produces haploid cells: cells with a single "dose" of genes or chromosomes. But if this separation does not happen, chromosome number is not halved and the sex cells produced will have the same diploid number as ordinary body cells. This failure is called non-disjunction. If successful fertilization happens

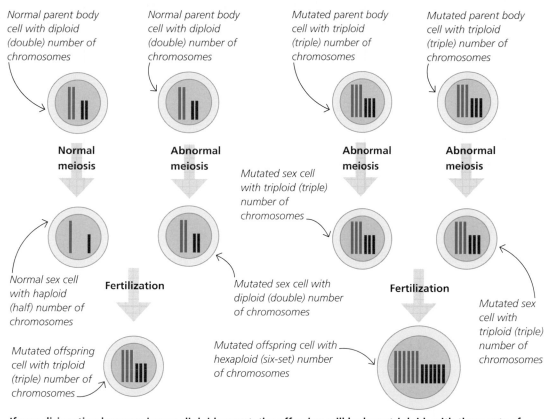

If non-disjunction happens in one diploid parent, the offspring will be born triploid: with three sets of chromosomes. An odd number of chromosomes cannot be halved during normal meiosis, so such individuals are infertile, unless non-disjunction happens again to produce hexaploids.

after that, the result will be offspring with more than two sets of chromosomes.

Extraordinarily, non-disjunction seems to be rife in certain organisms: many plants seem to do it as a matter of routine. We know this because around 25 per cent of plant species have cells packed with multiple sets of chromosomes. They are breaking the "diploid rule" that body cells have two sets only. This kind of mutation is called polyploidy, and specific terms are used for different numbers of sets. Triploid organisms have three sets, tetraploids have four, hexaploids have six, and octoploids have eight. But why does this happen so frequently? Depending upon the type of organism, polyploids can to do very well, sometimes much better than ordinary diploid ones. Perhaps this is because the

extra doses of genes help to provide additional masks against harmful recessive alleles, or improve the production of protein. Polyploidy is even common in some kinds of animals, especially fishes and amphibians. And, as we shall see in chapter 10, it can be a powerful driving force of evolution. But for other kinds of organisms, it appears to prevent normal development: polyploid human embryos invariably lead to miscarriage.

For those organisms in which polyploidy works, logistics would suggest that only even-numbered polyploids would be fertile. This is because normal meiosis involves halving chromosome number, which impossible to achieve with an odd number, for instance, in a triploid cell with three sets.

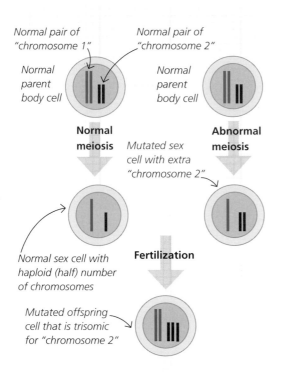

Normal pair of "chromosome 1"

Normal pair of "chromosome 2"

Normal parent body cell

Normal parent body cell

Normal meiosis

Mutated sex cell with extra "chromosome 2"

Abnormal meiosis

Normal sex cell with haploid (half) number of chromosomes

Fertilization

Mutated offspring cell that is trisomic for "chromosome 2"

Non-disjunction involving just one chromosome is called aneuploidy. Specifically, an offspring ends up with three "doses" of a particular chromosome, and is said to have trisomy with respect to that chromosome.

Aneuploidy

Sometimes non-disjunction happens, but it only affects one chromosome, rather than the entire package of chromosomes; this condition is called aneuploidy. Instead of multiplying up complete sets of chromosomes, aneuploidy results in an extra chromosome, or a loss of one. Unlike polyploids, many aneuploids cause harm. Some chromosomes are more vulnerable to non-disjunction than others. In humans, in about one in 1,000 eggs, chromosome number 21 fails to separate properly. Instead of producing two cells each with a single dose of chromosome 21, meiosis produces one with two and another with none. The cell with none

of the critical chromosome 21 genes cannot survive, but the other can be fertilized to produce an embryo with three doses. This condition, described as trisomy for chromosome 21, is commonly known as Down's syndrome. Individuals with the condition have 47 chromosomes rather than the usual 46. Other human chromosomes can suffer aneuploidy, resulting in a range of genetic abnormalities. Where this happens with sex chromosomes, arrangements such as XXY or X can occur.

Structural Chromosome Mutations

Some chromosome mutations happen in more subtle ways. Although the chromosome number looks normal, closer scrutiny of the DNA bundles reveals that pieces of chromosomes have joined together abnormally. This can happen because normal meiosis involves physical pairing of homologous chromosome partners. As we saw in chapter 5, this is necessary to mix genes by uncoupling ones that are linked on the same chromosomes. But faults in this crossing-over process can lead to structural problems. After prolonged crossing over, when chromosomes separate, sections of chromosomes can stay joined, so the break does not result in an even distribution of genes. A piece of one chromosome may stubbornly cling to another. Sections of chromosomes can even twist round, while remaining in their proper positions, and this also results in problems with development. These kinds of subtle mutations suggest that the proper position of genes on chromosomes is just as important as what the genes can do.

About 3 per cent of Down's syndrome cases are caused when a long extra section of chromosome number 21 is joined to chromosome number 14. This condition is called a translocation. It results in cells with 46 bundles of DNA, the normal human chromosome number, but they still carry the extra genes associated with the disorder. In this case, the condition results from a structural problem.

Chapter 10
EVOLUTION

Genes in Populations

Evolution happens at the level of populations, rather than individuals. Over many generations, the genetic makeup of a population changes as the group evolves new characteristics.

Change is a big part of life. Individuals switch behaviour in response to cues around them, such as a plant bending to face sunlight or an animal moving closer to food. And slower, more long-term changes happen as things grow older. Change through time follows every living thing from birth to death.

However, while the bodies of living things do not last for ever, their genes, at least potentially, can. Genes are copied in bodies and passed down from parents to offspring. They mix and re-mix as organisms breed with each other, producing new combinations of genes in new bodies. We have seen how mixing through Mendel's laws of inheritance can make a characteristic skip a generation. This means that offspring are produced that look different from their parents. But the genes themselves are transmitted intact. In an infinitely large population, inheritance alone does not change the numbers or proportions of gene variations in a population. If half the individuals carry only the dominant version of a gene, "A", and half carry only the recessive version, "a", there will still be 50 per cent "A" and 50 per cent "a" alleles generations later, as long as mating is random and no other factors force changes to the proportions.

In reality, additional factors do change these proportions. These factors can make one version of a gene predominate over another or drive others to extinction. Gene mutation can even generate entirely new genes: this is the source of genetic variation.

The ancient Greek philosopher Plato (428–348 BCE) believed that variations were imperfections of an ideal "type".

When the genetic makeup of a population changes in this way, we call it evolution. And the key to understanding it is to think in terms of populations, not individuals.

Thinking in Population Terms

Before Charles Darwin explained how evolution could happen, biology was stifled by a misunderstanding that diverted everyone's attention away from the true nature of living populations, and it had started more than two thousand years earlier with the Greek philosopher Plato. Plato regarded our everyday experiences of forms as being imperfect versions of their abstract essences. Shapes, objects, living things in the real world (so Plato said) are deviations from the perfect essences that exist only in theory. The notion comes from a strictly geometric view of the world.

A perfect square, for instance, exists only in the imagination, and any attempt to draw one would introduce imperfections. Naturalists adopted the Platonic view in their studies of animals and plants. For them, there was an ideal Garden Snail, and any variations in its shell shape or colour were deviations from "type". This meant that species were fixed objects of creation.

For Darwin, and all modern evolutionary biologists, there is no such thing as an ideal "type". Rather, every species consists of individuals that naturally vary, and any description of the characteristics of a species needs to take any such variation into account. Without variation, there can be no evolution. Evolution by mutation creates new genes, and so new varieties. Then, by various means, evolution makes the proportions of genes change down through generations.

Charles Darwin (1809–1882) had a different view from Plato about the nature of life. He believed that there was no ideal type, even in theory, and that natural variation was necessary for evolutionary change.

Population Genetics

By the mid-nineteenth century, Charles Darwin was building his theory of evolution. This stated that, among the wide variation of living things inhabiting an environment, the individuals that were best adapted to the conditions in that environment survived and reproduced more successfully to pass on their inherited characteristics. At the same time, Gregor Mendel was carrying out his experiments in cross-breeding varieties of the garden pea. Mendel concluded that inheritance of characteristics was caused by passing on the particles that we call genes. The varieties of genes, called alleles, came in dominant and recessive forms. When a recessive allele was masked by a dominant one, its effects skipped generations and remained hidden in many individuals within a population. When Mendel's work was rediscovered at the start of the 1900s, biologists began to focus on bringing Mendelian inheritance and Darwinian evolution together in an all-encompassing theory that became known as the "modern synthesis". But as biologists studied the new field of population genetics, many insisted that there were big problems in reconciling the two ideas.

In particular, it was not clear how genetic variety could be maintained in a population.

Some thought that the underlying processes would automatically make dominant traits increase, while recessive ones would diminish. But this was not what biologists were seeing in the natural world, where all sorts of varieties apparently persisted. The solution was a mathematical one, and it came, independently, to two researchers in 1908: British mathematician G H Hardy and German physician Wilhelm Weinberg. It became known as the Hardy–Weinberg Principle.

An Equilibrium of Alleles

Imagine you have a large population of fruit flies that, for a particular gene, contain a mixture of insects with dominant and recessive traits. A simple example is a gene that controls the development of their wings. Normal wings are determined by a dominant allele, which we will call it "N". A recessive allele, "n", makes the wings short and useless for flying: a condition called vestigial-winged. Since the normal allele is dominant, some normal-winged flies in the population could be carrying the recessive allele, meaning that their genotypes (genetic combinations) would be NN or Nn. But all vestigial-winged flies would have to be nn. Now imagine that we are setting up a population that contains equal numbers of N and n

British mathematician G H Hardy (1877–1947) was one of the authors of the Hardy–Weinberg principle. The principle states that allele frequencies within a population of living things will stay constant from one generation to another unless they are acted on by evolutionary influences.

alleles. Such a population might contain genotypes in the following proportions: 25 per cent NN, 50 per cent Nn and 25 per cent nn. Since some recessive alleles are "hidden" in the Nn individuals, only a quarter of the population of flies will have vestigial wings.

Now the flies start to mate randomly. Every individual coupling obeys Mendel's laws. For instance, if two Nn flies mated, a quarter of their offspring would have vestigial wings. If NN crossed with nn, none of them would. And if Nn crossed with nn, half of them would be vestigial. Hardy and Weinberg predicted that, with all the random possibilities considered together, even after several generations the proportions of N and n alleles would stay exactly the same, as would the proportions of genotypes (NN, Nn and nn). With 50 per cent N and 50 per cent n, those

would be the proportions of genotypes that we would expect to persist given all the random mating. In other words, the genetic makeup of the population would stay around this so-called equilibrium from one generation to the next.

The Hardy–Weinberg principle predicts that the genetic variation across a large population would be maintained like this as long as inheritance alone (and not evolution) was taking place. It is true for all kinds of genetic variation, including ones that follow more complex patterns of inheritance, such as sex linkage. The idea became the linchpin of the new field of population genetics and showed that Mendel's laws were perfectly compatible with the persistence of genetic variation in entire populations. It also showed that inheritance alone does not lead to a change in the genetic makeup of a population. But what would?

Inheritance alone does not explain change in the genetic makeup of a randomly mating population (except through genetic drift – see page 135). Here, 16 fruit flies collectively contain a total of 32 wing genes (two each), half of which are normal and half of which encode for recessive vestigial wings. After many generations, the genetic makeup stays in equilibrium, even though flies die and new ones are born: with a 50–50 mix of alleles that equilibrium always contains a quarter normal-winged non-carrier flies (NN), half carriers (Nn) and a quarter vestigial-winged flies (nn). With different proportions of alleles, different proportions of genotypes would be predicted.

N = allele for normal wings
n = allele for vestigial wings

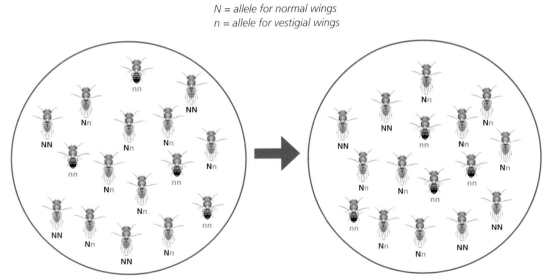

Original population of fruit flies

Population of fruit flies several generations later

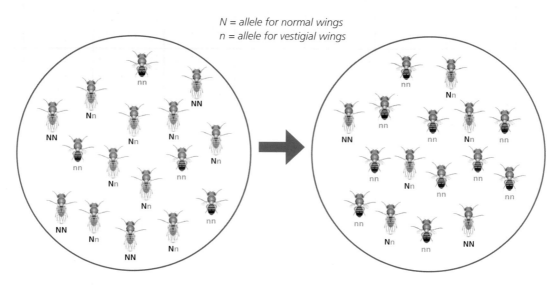

N = allele for normal wings
n = allele for vestigial wings

Original population of fruit flies

Population of fruit flies several generations later

Evolution takes place when the genetic makeup of a population changes. Here the population of fruit flies has evolved to produce a greater abundance of vestigial-winged flies. It could be that conditions have changed to make these variants breed better. Note that the frequency (proportions) of alleles have changed: now only 25 per cent of the alleles are N, while 75 per cent are n.

Evolutionary Change

The Hardy-Weinberg principle, and the predictions it makes, gives us a stable situation with which we can compare evolving populations. It therefore helps us to focus on what causes populations to change through evolution.

Mutation is one process that will throw a Hardy-Weinberg equilibrium off-balance. If any one of the fruit flies in our population had suffered a mutation of their wing gene, this would have added a new, third, allele to the mix, and so shifted the proportions of N and n. And imagine that this new allele made the flies better at growing and reproducing. Such an allele could now spread at the expense of the others. Now the proportion of alleles would be off-balance as the proportion of the new allele rose, and proportions of old ones diminished. Over time, the genetic makeup of the

population would change: evolution would have taken place.

In reality, few natural populations are in exact equilibrium in the way the Hardy-Weinberg equilibrium predicts. In most populations, there is some evolutionary change. Although evolutionary change, as we will see, can work within a short time, much of it might be very slow. But over a far longer time frame, involving hundreds or thousands of generations, even this effect can be very clear indeed. And one of the biggest driving forces of evolution, leading to the emergence of adaptation, is the process of natural selection. This process was was explained at length by Charles Darwin in his great work *On the Origin of Species by Means of Natural Selection, or the Preservation of Favoured Races in the Struggle for Life*, published in 1859.

How Evolution Works

Evolution is said to have happened when the proportions of genes in a population change from generation to generation. This is caused by a combination of mutation and selection.

Every life form on Earth is related to every other life form. Such is the inevitable conclusion that must be drawn from the fact that all living things are similar, at least, at the levels of cells and molecules, including DNA. There are genes that control the most fundamental processes of life, such as making proteins and generating energy, that are practically the same in organisms as different as bacteria, oak trees and elephants. Overwhelmingly, all the evidence points to all life having descended from a single common ancestor.

Today we understand better than ever how life has evolved over countless generations to produce the biodiversity around us. Not only can we explain it in terms of the behaviour of genes, but we can even watch it happening in fast-growing organisms in the laboratory. And by looking at the evidence in living bodies and fossil remains, we can work out how changes that can be observed through generations add up over millions of years to produce different kinds of organisms.

What Does Evolution Mean?

Whenever the genetic makeup of a population changes from generation to generation, evolution has happened. Technically, this is determined by changes in the proportions, or frequencies, of particular alleles. An evolving population is one that breaks free from an equilibrium (the so-called Hardy–Weinberg equilibrium) that would otherwise

Most humans, like most mammals, lose their ability to digest milk when they grow up into adulthood. But dairy consumption increased with farming, and at some point in the last 10,000 years, a new allele arose and spread through Europe that allowed people to digest lactose into adulthood.

6,500 years ago Well-developed dairy economy established in central Europe.

11–10,000 years ago Neolithic culture develops in the Middle East. This is the start of agriculture and possibly the domestication of dairy animals.

8,000 years ago Neolithic reaches the Balkans.

8,400 years ago Neolithic spreads to Greece.

keep the genetic variety the same. Evolution is happening, for instance, when a gene that encodes for a milk-digesting enzyme becomes more common, at the expense of its non-milk sugar digesting variant. Most adults around the world cannot digest the sugar found in milk, but as dairy farming spread during humankind's history, so the allele for digesting milk sugar increased in frequency, too. Much of evolution involves subtle changes like this, taking place over the course of a few generations, but given longer, the effects of evolution build up.

In reality, of course, evolution brings about changes in many different genes. The additive effects of all these changes lead to the emergence of characteristics that are so different that the organisms that carry them could be classified as different species. These longer-term effects, called macroevolution, are explained in the next section, but here we begin with smaller evolutionary changes that happen within single species: microevolution.

Adaptation

As well as explaining population change through generations, any evolutionary theory also has to explain adaptation. Adaptations are features that make organisms especially suited to the environment in which they are found. An obvious feature of biological diversity is that living things show adaptations to their surroundings: they are not randomly distributed. For instance, organisms found in freezing polar habitats have high tolerance for cold; some may even have built-in anti-freeze chemicals. Animals, whether hunters or hunted. may have a white coat to be camouflaged against snow or ice.

Much genetic change, such as mutation, is entirely random, and cannot, on its own, be used to account for the emergence of specific adaptations. Charles Darwin's theory of evolution by natural selection became a powerful explanation for adaptive changes.

Evolution

How Darwin Explained Evolution

In ancient Greece, philosophers such as Empedocles proposed that living things descended from others, but for centuries this idea was rejected by most naturalists in favour of a system of fixed types that were created according to a divine plan. It was not until the 19th century that scientists came closer to understanding biology in terms of physical laws. In 1809, the French naturalist Jean-Baptiste Lamarck proposed that life forms evolved to become more complex, and did so by inheriting features that became overdeveloped through use, or underdeveloped by disuse. In Lamarck's view, the long neck of the giraffe evolved because its ancestors stretched to reach the highest foliage, and then passed these stretched necks on to their offspring. Today, we know that genes determining neck length cannot be influenced in this way. But Lamarck created one of the first serious evolutionary theories, and his theory did at least explain adaptations.

Charles Darwin developed his own theory of evolution over decades. Crucially, his theory depended upon the recognition that species were highly variable, and not fixed to types. Darwin envisaged that some of these varieties were more successful than others. These were the ones that survived better to produce more offspring. If the variation was inherited, then the more successful varieties would pass on their favourable characteristics. Compared with Lamarckism, Darwinism began with an ancestral population of giraffes with slightly different necks, but the longest-necked ones reached more food, became better nourished and had more babies. The key was inherited variety, not use and disuse.

Darwin published his work in 1858, in a joint presentation of the Linnean Society of London, with a fellow naturalist called Alfred Russell Wallace, who independently came up with the same idea. Darwin greatly elaborated on his theory in his book *On the Origin of Species*. He called it natural selection because the environment of living things was ultimately selecting the most favourable varieties.

Natural selection predicts that individuals with genes that give them an advantage over their competitors will survive longer and produce more offspring. As more of these "fitter" individuals predominate, the species becomes adapted to its surroundings. If the environment changes, different individuals may then have the selective advantage so evolution will produce new adaptations.

1. There is genetic variation in the population of insects (due to mutation).

2. Some individuals survive better than others, depending upon their characteristics. In this case, insects with alleles that make them green escape predation because they are camouflaged.
The bird can see the pink and orange insects more clearly than the green ones, so only the pink and orange ones get eaten.

3. The ones that survive also breed, passing on their favourable alleles. The green insects therefore breed to produce more green insects, so most insects end up being green.

The Growth of the Modern Synthesis

At the turn of the 20th century, some biologists were still doubting whether Darwin's natural selection could adequately explain evolution, just as some were doubting the rediscovered work of Gregor Mendel in explaining inheritance. But, as the emerging science of genetics began to confirm the roles of genes and chromosomes, so Darwinism and Mendelism began to converge into a new view of evolutionary biology: the so-called "modern synthesis". After Hardy and Weinberg showed that Mendelian inheritance made sense in natural populations, other biologists began to study population genetics in more detail. By 1930, British geneticists such as Ronald Fisher and JBS Haldane were showing that natural selection could change frequencies of alleles and so cause evolution in the way Darwin had described. At the same time, populations were found to contain much more genetic variation than previously thought: more than enough for evolution to happen by changes in allele frequencies. The modern synthesis of evolution had determined that evolution, through changing allele frequency, could be caused by factors such as mutation and selection.

Evolution by Mutation

As we saw in chapter 9, mutation is the ultimate source of all new variation. New alleles, and entirely new kinds of genes, are created by DNA-copying errors. Some of these new genes are harmful, some are beneficial, while others may have no effect on wellbeing at all. But all will add to the genetic variety of a population.

Unlike selection, however, mutation is random, so it alone cannot account for the routine appearance of characteristics that make living things adapted to their environment.

Natural selection, where the "fittest" species in an environment survive over "unfit ones", offers the only workable explanation for evolutionary processes.

Evolution by Natural Selection

Only evolution by natural selection provides an explanation for evolutionary change that results in adaptation: where organisms are selected to "fit" the environment around them. Natural selection relies on the genetic variety produced by mutation. This variety means that the individuals in a population will vary in fitness. Some survive better and longer to reproduce more. This means that they are more likely to pass on the genes that helped them outperform their competitors. But, critically, the kinds of genes involved depend upon the environment. If the environment changes, then different genes may give the selective advantage. Either way, subsequent generations will inherit the appropriate genes, and the inherited characteristics of a population will gradually adapt to their surroundings.

This process can be observed in action. Warfarin is a drug that is widely prescribed to minimize the risk of strokes in patients with high-risk conditions, such heart disorders. But it is also used as a pesticide to control rats. It works by blocking an enzyme involved in the blood-clotting process. The effects in rodents are fatal internal haemorrhages. It was introduced as a pesticide in 1950, and has been used ever since because of its low toxicity to other animals. But eight years later, brown rats resistant to the drug appeared in Scotland. They carried a mutation that prevented the drug from interfering with the blood-clotting process. Since then, different mutations in mice and rats with similar effects have materialized all around the world. Warfarin is the environmental factor that is selecting rodents with the resistance. This means that, where warfarin is being used, the resistant rodents survive, but the non-resistant ones die. As a result, natural selection is driving the spread of warfarin resistance.

Outbreaks of pesticide resistance in the UK in rodents (shown by the orange areas) have occurred wherever the pesticides have been used, as rodents acquire resistance through mutation and natural selection. In Britain, separate outbreaks in Scotland, Wales and England are due to separate, independent mutations.

Types of Natural Selection

The spread of warfarin resistance in rodents is an example of directional selection: it leads to a rise in the levels of a particular allele, in this case the allele that conferred resistance. The proportion of the "normal" non-resistant version of the allele therefore diminishes. But there is a twist to the story. How does the situation play out in places where warfarin is not being used? Bizarrely, the non-resistant rodents then have the advantage. Although warfarin resistance helps rats survive the drug, it comes at a cost: they need more vitamin K in their diet, a nutrient that is also involved in the complex blood-clotting process. Where warfarin is used as a pesticide, this disadvantage is more than offset by the protection against the poison. But in the absence of warfarin, it is enough to shift the balance and the non-resistant rodents do better. As a result, selection will then tend to work against the extremes and in favour of "normal" rodents: this is called stabilizing selection.

In fact, most times it is stabilizing selection that is at work in a population. Directional selection, in comparison, is usually more associated with a change in the environment. Stabilizing selection keeps the status quo by getting rid of the outliers of a characteristic's range. In the past, natural selection has worked to drive the evolution of ornamental plumage in many kinds of birds because males with colourful displays are better able to attract females. This so-called sexual selection explains why peacocks have evolved long colourful trains. But in wild peafowl, the train is a liability when it comes to escaping predators: fully-plumed males are weaker fliers. Stabilizing selection will work against the alleles associated with the shortest trains, because such birds would not get a female. But if the trains are too long, the alleles are targeted by predators.

A third kind of selection works in favour of several genetic variants at the same time: this is disruptive selection. Populations of brown-lipped

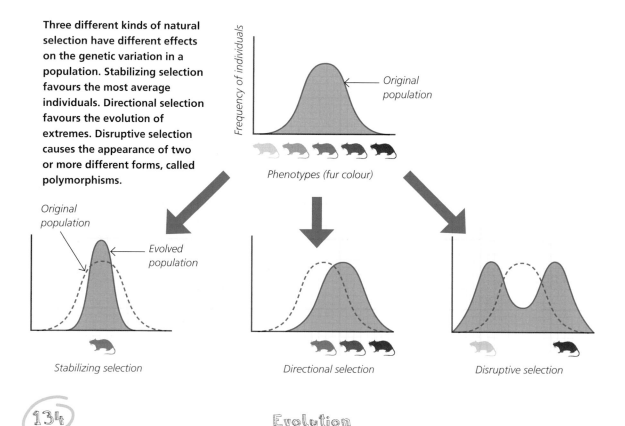

Three different kinds of natural selection have different effects on the genetic variation in a population. Stabilizing selection favours the most average individuals. Directional selection favours the evolution of extremes. Disruptive selection causes the appearance of two or more different forms, called polymorphisms.

Frequency of individuals

Original population

Phenotypes (fur colour)

Original population

Evolved population

Stabilizing selection

Directional selection

Disruptive selection

Evolution on islands may happen fast because of the effect of random genetic drift. Founder populations are likely to be very small, made up of just a few individuals that have dispersed there from a distance mainland. Island species, such as the dodo (right) or giant tortoises, are often very different from their closest mainland relatives.

snails, a common European species, contain a mixture of animals with dark brown shells, yellow shells or stripy shells. Disruptive selection manages to maintain the necessary genetic variety because each of the three colours is camouflaged from predators, mainly thrushes, in three different habitats: dark woodland, open heath or sun-dappled hedgerows. Sometimes the disruptive effects are less visual. Sickle cell disease is caused by a mutation that deforms red blood cells (see chapter 3). In countries with inadequate medical care, mainly in the tropics, sufferers may not survive long enough to reproduce. But in such places, carriers of the harmful recessive allele do just as well as people who lack the allele altogether. The reason is that the carrier "trait" makes them more resistant to malaria, a disease transmitted by blood-sucking mosquitoes. As a result, disruptive selection in the tropics favours both carriers and people with normal blood.

Random Genetic Drift

Imagine we had a population of fruit flies that, once again, is made up of a mixture of normal and vestigial-winged insects. As we saw in the previous section, inheritance alone would not change the frequency of wing alleles from generation to generation. Only evolutionary change, such as mutation or selection can do that. But one kind of evolutionary change has a subtle effect. This is called random genetic drift. And, as the name implies, its random effects do not result in adaptation.

If the population of flies is very large, and all the insects are breeding randomly, the allele frequencies will, indeed, stay much the same, even after several generations have passed. But now consider what might happen if the population were very small, consisting of just six flies. Now, the likelihood of genetic change happening entirely by chance rises. This is because, although the flies are mating randomly, it is more likely that the alleles that get passed on are not a good representative of the parental variety. Entirely by chance, for instance, it is possible that, among just six flies, a disproportionate number of vestigial-winged flies might interbreed, leading to a greater fraction of alleles for vestigial wings being passed on. Each new generation carries, effectively, a sample of alleles from the parental population that produced it. Genetic drift is this kind of chance slippage in allele frequency. In reality, it will have an effect in any population, but its effects are more pronounced in populations that are very small.

Random genetic drift helps to explain why naturally small populations, such as those isolated on tiny islands, tend to evolve more quickly than larger mainland populations.

When a worker honey bee stings a human to protect her hive, she loses her life. But she is helping to secure the safety of her mother, the hive's breeding queen, who is continually laying eggs that contain worker bee genes.

Selfish Genes

If individuals live and die in the evolutionary struggle for existence, it may seem that individual organisms are the prime responders to evolution's drive of natural selection. But in 1976, British biologist Richard Dawkins published a book, *The Selfish Gene*, in which he challenged that idea. Dawkins proposed, instead, that the gene is the ultimate "unit" of evolutionary selection. He suggested that self-replicating genes are acted upon by natural selection, rather than living bodies, which are nothing more than vehicles for transmitting them. Genes worked together inside a body, but, ultimately, each has a selfish drive because of its tendency to replicate. Individuals carrying genes with the most effective drive are the ones that survive and breed better.

The selfish gene idea was devised as a way of debunking the popular notion that entire populations or groups of living things could be naturally selected. Dawkins argued that a gene-centred view of selection served to focus on the unit that really mattered: the genes themselves. This also helped to explain apparently altruistic behaviour. Worker bees, for instance, are sterile and toil solely for the benefit of the egg-laying queen. In evolutionary terms, this sacrifice seems not to make sense. How could natural selection favour non-breeding workers? In fact, because workers are daughters of the queen, they will share her genes. This means that the genes that control the toiling behaviour of the sterile workers are, indirectly, ensuring that their replicas are continually passed on from the body of the queen.

Evolution

New Species From Old

Large evolutionary changes can lead to the appearances of individuals that are so different from others that they fail to interbreed with them. These emerging kinds of living things are new species.

In the 1920s and 1930s, as biologists and mathematicians grappled to understand the behaviour of genes in populations, the discussion turned to the way entirely new species of organisms can emerge from the diversity of living things. Although Darwin called his book *On the Origin of Species*, he concentrated on how natural selection would drive evolutionary change. He avoided the tricky problem of how this change could break free from one kind of species and lead on to another.

We are familiar with species in our everyday lives as kinds of organisms that appear to be distinctly different from others, even though they may share certain characteristics and be obviously related to them. The birds that visit a garden feeder can usually be identified easily as species of finches, tits, and so on. There is a discontinuity between one species and another: not only do they look different, but they may behave differently too. But if evolution is such a gradual change involving shifts in predominance of varieties of genes, how can one species transform into another: what would make this kind of cross-species evolution, called macroevolution, happen?

In order to answer that question, biologists first needed to ask themselves a more fundamental question: what defines a species in the first place?

Natural selection can produce a wide range of species with very different characteristics.

Defining Species

In 1937, Soviet-born geneticist Theodosius Dobzhansky published a book entitled *Genetics and the Origin of Species*, which combined many aspects of evolutionary theory at the time. Dobzhansky had emigrated to America, where he began to work on fruit flies with Thomas H Morgan, who had confirmed that genes are carried on chromosomes. But Dobzhansky was as interested in understanding populations of living things in the wild as he was in laboratory work.

Willow tit

Marsh tit

The willwo tit and marsh tit are two closely related birds of European woodland. Until 1897, they were thought to be the same species, but their distinctive calls – the most reliable way to separate them – mean they behave as separate species.

His book brought together ideas about mutation and selection in natural populations, supported by evidence from experience and observations in the field, but also covered a related issue that became key: the way that species could evolve by isolation. Populations could split into different species when the individuals involved stopped interbreeding.

A few years after the publication of Dobzhansky's book, German-born biologist Ernst Mayr seized upon the notion of isolation as a way of defining species: a population of individuals that could interbreed among themselves, but could not interbreed with other such groups. Species were defined as groups that were reproductively isolated from others. Reproductive isolation makes species genetically distinct because their genes do not mix with the genes of others. In other words, their gene pools are different and incompatible. Mayr's so-called "biological species concept" was adopted by naturalists working in all fields, and is still popular today. Two kinds of organism may look and even behave very similarly, but if they fail to interbreed, they must be classified as different species according to Mayr's definition. For example, the willow tit and marsh tit are two similar-looking birds that flit about in woodland habitats in search of insect food. But their calls are very different, meaning that they do not court one another, and so never interbreed.

Geographical Barriers

With a solid idea of what a species was, biologists could properly address the issue of how one species can evolve from another: a process called speciation. The work of Dobzhansky and Mayr laid the foundations of our understanding of macroevolution. In order for it to happen, evolution has to create some sort of reproductive barrier between individuals that stops the flow of genes between them: a so-called reproductive isolation mechanism. The easiest way to envisage this happening is with a physical geographical barrier.

Organisms do not live in a permanent, unchanging world. Darwin had already shown that shifts in environment and habitat can create different conditions under which new varieties can prosper by natural selection. But in the long geological past, over millions of years and countless generations, the surface of the Earth has changed very dramatically indeed. Modern science has abundant evidence that continents have moved position, lands have fractured and collided, mountains have risen and valleys have dropped. In the Jurassic age of the dinosaurs, South America, Africa, Madagascar, India and Antarctica had been joined together for hundreds of millions of years, but they began to separate and have been moving ever since. When India collided with Asia, it pushed up the Himalayas. And as the South American Andes rose, they flooded the Amazon basin. All these changes had a profound effect on living things: as barriers rise and fall, so populations fragment or join.

A population of a single species can break up through the emergence of geographic barriers, such as new mountain ranges or rivers. For instance, a species that is adapted to low elevations will not cross the high peaks of a mountain. This means that the populations on either side of the new barrier will each evolve in their independent ways. Sometimes the conditions will be very different: perhaps the mountains create a rain shadow so one population is exposed to a rainy habitat and the other to an arid one. This will make natural selection drive evolution in different ways. Or perhaps, the populations evolve differently by chance: through random genetic drift or mutation. In either case, after many generations, the two populations could evolve to be so different that, were they to meet, they would not be capable of interbreeding. They would have evolved to be reproductively isolated from one another: they would be different species.

Just a few million years ago, North and South America were separate continents, but they joined across the Central American land bridge as continents moved. This separated populations of marine animals in the Pacific and Caribbean areas, leading many, such as porkfishes, to evolve into separate species. Comparison of genes in these species pairs provides evidence for how this happened.

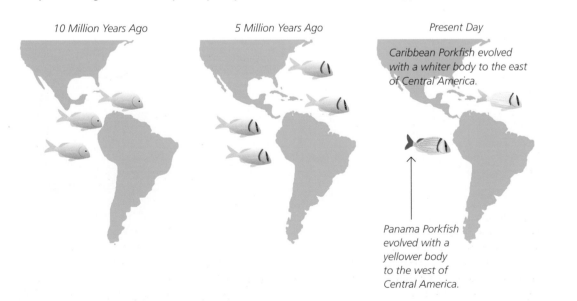

10 Million Years Ago

5 Million Years Ago

Present Day

Caribbean Porkfish evolved with a whiter body to the east of Central America.

Panama Porkfish evolved with a yellower body to the west of Central America.

In all cases of speciation, reproductive isolation must evolve in order to prevent newly emerging species from interbreeding with old ones. Unlike allopatric speciation, sympatric speciation achieves this without a geographic barrier.

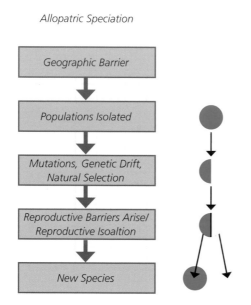

Allopatric Speciation

Geographic Barrier

Populations Isolated

Mutations, Genetic Drift, Natural Selection

Reproductive Barriers Arise/ Reproductive Isoaltion

New Species

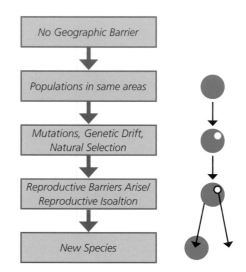

Sympatric Speciation

No Geographic Barrier

Populations in same areas

Mutations, Genetic Drift, Natural Selection

Reproductive Barriers Arise/ Reproductive Isoaltion

New Species

Reinforcing the Difference

Dobzhansky and Mayr championed the idea that many new species emerged due to geographic barriers. This is is called allopatric speciation, from the Greek *allos* meaning "other" and *patris* meaning "fatherland". The idea has been supported by the results of laboratory experiments with fruit flies, in which groups of insects were isolated and exposed to different conditions. After many generations, when the isolated groups were recombined, the insects preferred to keep mating with their own kind. Evidence also comes from studying the natural distribution of species in the wild. Closely related pairs of species are often geographically aligned on either side of a geographic barrier, and the most closely allied ones may even exhibit a hybrid zone between them, suggesting that speciation has not yet run to complete incompatibility.

The extent to which species stay separate after coming together depends on the genetic differences that have built up over time. With sufficient genetic change, populations can be so different that no individuals would attempt to interbreed. Research indicates that hybrids are more common than once thought, but many biologists think that hybridization itself reinforces the speciation process. As genes and chromosomes evolve in populations that were once separated, they may become more or less incompatible when they mix together after a barrier between them broke down. For instance, recombining differently evolved genetic material could produce a hybrid that is infertile. If this happens, natural selection would continue to work against these "unfit" hybrids. Over time, species would diverge even more at their point of contact, as individuals that only interbreed among their own kind succeed in passing on their genes.

Evolution

No Geographical Barriers

Allopatric speciation satisfied so many examples of evolution that it appeared to be the only good explanation. Ernst Mayr, in particular, insisted that no other method would work. However, in the 1940s and 1950s, a controversial idea about evolution was brewing, culminating with a scientific paper published in 1966 by the British biologist John Maynard Smith. In the paper, he argued that new species could appear within the distribution of old ones without geographic interruptions to gene flow. It became known as sympatric speciation, meaning "together" and "fatherland".

Maynard Smith was suggesting that a population that carried sufficient genetic variation could undergo divergence from within its gene pool (see diagram on page 140). His idea was that natural selection could favour two or more variants at the same time, so the divergence could be driven by disruptive selection.

Darwin's Finches

The most famous example of Darwin's illustration of evolution by natural selection came in the form of a group of finches found only on the Galápagos Islands in the Pacific Ocean. Darwin visited the islands in 1835 as part of a round-the-world voyage. Contrary to legend, the bird specimens he collected there at the time did not provide him with a "eureka" moment. But back in London two years later, he saw how they would play a key role in his growing theory of evolution by natural selection. His ornithologist friend John Gould identified them all as divergent kinds of finches. Darwin had thought they were a mixture of blackbirds and grosbeaks, but they shared characteristics that made them finches, while being sufficiently different to be obvious species. In particular, their beaks were divergent, ranging from a needle-like form usually associated with insect-eaters to a chunky form seen in seed-eaters.

Darwin's finches are not distributed randomly: the different species are found on specific islands. In *Origin of Species*, Darwin demonstrated how they had descended from a common ancestor that had arrived on the islands from nearby South America, and that they had evolved by natural selection to exploit different kinds of food found on each island. Since then, biologists have extended our understanding and today Darwin's finches are among the most studied birds anywhere in the world. In 1947, British ornithologist David Lack published a landmark study, and since the 1970s, British biologists Peter and Rosemary Grant have carried out decades of painstaking field research.

The results of these studies have shown that Darwin's finches are, indeed, genetically related to finch-like birds in South America and the Caribbean, and that the initial colonization diverged by allopatric speciation. But on some islands, speciation also happened sympatrically. Disruptive evolution favoured extreme bill sizes, which led to the evolution of exaggerated forms without geographical separation.

Finches isolated on separate Galápagos islands diverged by allopatric speciation, but some found on the same island did so by sympatric speciation. These are Darwin's original drawings.

1. *Geospiza magnirostris*
2. *Geospiza fortis*
3. *Geospiza parvula*
4. *Certhidea olivasea*

Evolution by Multiple Chromosomes

In chapter 9, we saw how dramatic changes to the genetic makeup of an organism can happen when chromosome mutations change chromosome number. Specifically, if chromosome pairs fail to separate when sex cells are made, multiple sets of chromosomes can end up in subsequent generations: a condition called polyploidy.

In some organisms, such as many plants and some species of fish, polyploidy is a regular occurrence. In these cases, it can effectively cause reproductive isolation within a single generation. Although different individuals sharing the mutation may interbreed among themselves, their chromosomal differences may bar them from breeding with individuals that carry the chromosomal makeup of their parents. There are numerous examples where new species have emerged in this way within just one or a very few generations. This creates a speciation process that is effectively instantaneous.

Goatsbeards *(Tragopogon)* are members of the sunflower family, mainly found in Eurasia. Some species have been introduced to North America, including *Tragopogon dubius* (top) and *Tragopogon pratensis* (middle). In the 1950s, a new species was found in the states of Idaho and Washington. It had emerged, by polyploidy, through hybridization of *T. dubius* and *T. pratensis*. The new species, named *Tragopogon miscellus* (bottom), is a tetraploid: a species with four sets of chromosomes.

Evolution Through Deep Time

As evolution happens across millions or billions of years, the divergence associated with new species builds up so that organisms become ever more different. Over geological time, entire new groups of organisms evolve as biodiversity increases.

As conditions on the Earth change across eons of time, species come and species go. A species of plant or animal may last a million or so years before becoming extinct or evolving into something very different. A few species are unusually resilient and seem to have persisted since the age of the dinosaurs or even before, while other species, especially fast-breeding kinds, change within centuries. But for as long as they last, species keep their identities and continue to interbreed to produce more of their kind.

We can compare the characteristics and genes of organisms alive today to work out their evolutionary history, building up a family tree of life that shows how new kinds branch from others. The species that are alive today are the very ends of the terminal twigs of this ancestral tree. In that respect, the evolution of new groups of organisms is the same as the evolution of new species, only multiplied up over much longer periods of time. Humans are more genetically similar to apes and monkeys than they are to birds or fishes, and we are even more distant from insects and plants. But ultimately all known life is more or less related because all living things share a single common ancestor.

Despite the wide range of species that are living and have lived on the planet, their roots can be traced back to a single common ancestor species.

Evolution Over Geological Time

The geological time frame spans the entire age of the Earth. That's just over 4.5 billion years. The ages of the rocks beneath our feet can be determined by several physical tests. The oldest fossils, about 3.5 billions years old, define the timescale of the evolution of life. Across these great stretches of time, groups of plants, animals, fungi and microbes have evolved from simple single-celled ancestors.

There are many more species alive today than there were at life's beginnings, which means that the evolutionary history of life on Earth takes the form of a branching tree. The continual process of speciation has ensured that the tree keeps branching over and over again. As the tree matures, new species diverge so far from their very first branching points that they end up being as different as ferns and roses, or worms and whales. This successive branching in evolution is called cladogenesis, and each new branch of life that emerges is called a clade. Clades begin as species, but over time can diverge into entirely new families, phyla or kingdoms of life.

But accumulated evolutionary change can also happen without branching. If allopatric and sympatric speciation does not happen in space, one species can completely change into another just through time. A population that persists over a million or so years without fragmenting is unlikely to be genetically the same at the end of that duration as it was at the beginning, and could well be so different that, if individuals from either end of the timescale were to meet, they might not be able to interbreed. This gradual change along a single branch of the tree of life is called anagenesis.

Evidence of Relationships

Organisms that are most closely related are most similar in terms of their genes. A range of techniques is used to work out similarities or differences in genes. By applying statistical

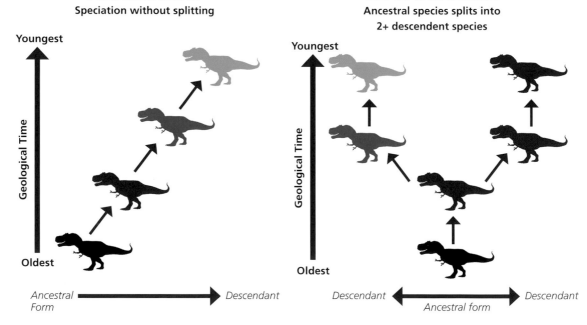

Over geological time, populations may accumulate genetic changes until they become a different species from their ancestors (anagenesis), or they may split into more than one species (cladogenesis).

Fishes, whales and some prehistoric marine reptiles have all evolved streamlined shapes with fins or flippers, because selection has favoured these features for living in water. In other words, their features have converged.

methods, the results of such comparisons can be used to build up the single most likely branching tree pattern that produced the diversity.

Of course, evidence from genes has not always been available to biologists. Before the emergence of molecular biology, naturalists had to rely on the evidence from anatomy or behaviour to sort out evolutionary relationships. In many cases, these studies were reliable, but there were pitfalls. In particular, it is possible for very distantly related species to evolve similar features because natural selection is acting on them in similar ways.

Evolutionary convergence can also happen at a molecular level, but on the whole DNA comparisons are more reliable, especially if many sections of DNA are compared simultaneously. It is not very likely that the same stretches of base sequences will emerge independently. Some genes or sections of DNA are not likely to be influenced by selection, and these are the ones that will often provide the most accurate indication of evolutionary relationships. The degree of difference is likely to have happened by mutations that have built up randomly over time, and this gives a good indication of the evolutionary "distance" between species. Studies of fossil species can rarely use evidence from genes because the DNA has usually degraded. But recent gene comparisons have uncovered new facts about the evolutionary relationships of modern species.

Genetic studies have confirmed that cetaceans are divided into two branches: the toothed cetaceans, such as dolphins, and the filter-feeding baleen cetaceans, such as blue whales and minke whales. This idea was previously supported by anatomical studies. The same studies indicated that cetaceans were a "sister" branch to a branch that carried cloven-hoofed mammals such as cattle. But in the 1990s, gene analysis showed that cetaceans had instead emerged from within the hoofed mammals, specifically paired with hippopotamuses.

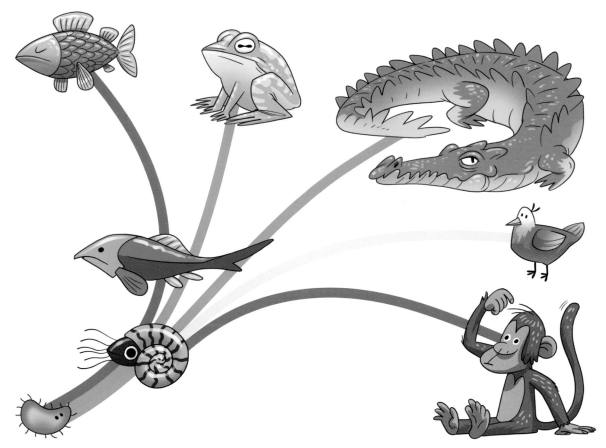

The tree of life, as revealed by comparing genes, is a reflection of how groups of living things have evolved and diverged through millions of years.

Constructing the Entire Tree of Life

Gene studies have helped to clarify the relationships among the main branches of the tree of life. Traditionally, organisms were split into animals and plants. The animals were thought to include many moving single-celled organisms, such as amoebas, while the plants were lumped with other "vegetative" things, such as fungi and bacteria. But microscopic studies of cells, and then analysis of genes, showed just how artificial this was. In the 1930s, bacteria were thought to be so different from everything else that they were given their own kingdom. Over subsequent decades, algae and fungi were separated too.

Today, the evidence from genes shows that there are many branches at the base of the tree of life, so much so that it looks more like a shrub than a tree. In the 1970s, some single-cells previously grouped with bacteria were found to be as genetically different from them as "true" bacteria were from people. The new group, called archaeans, included bizarre life forms that seemed survivors of a very distant prehistoric age. Some live today in boiling hot acid springs, or generate methane in the stomachs of herbivores. Meanwhile, gene studies were showing that fungi were more closely related to animals than to plants, and that a whole range of complex single-celled organisms were not very closely related to either.

Evolution

Molecular Clock

In 1968, the Japanese geneticist Motoo Kimura suggested that a great deal of genetic variation that accumulates over time has been produced randomly by mutation and drift . Aspects of this so-called neutral theory of molecular evolution remain controversial, but the idea that the base sequences of genes accumulate "neutral" mutations is supported by observation. Kimura went on to suggest that the build-up of these mutations would happen at a constant rate over long periods of time because they are not being influenced by selection. In this respect, genetic differences associated with this sort of change could be used as a clock to measure the passage of time. It meant that, at least in theory, geneticists could work out the time when different species diverged.

This molecular clock method is used today to estimate times of branching points on the phylogenetic tree of life. Although the clock alone can only be used to compare one branching point against another and cannot give absolute dates, it can be calibrated by different means. For instance, fossils, which have been dated by testing rocks, can be incorporated into the family tree near key ancestral branching points. By using fossils that are thought to be close to these points, the dates of divergence can be estimated.

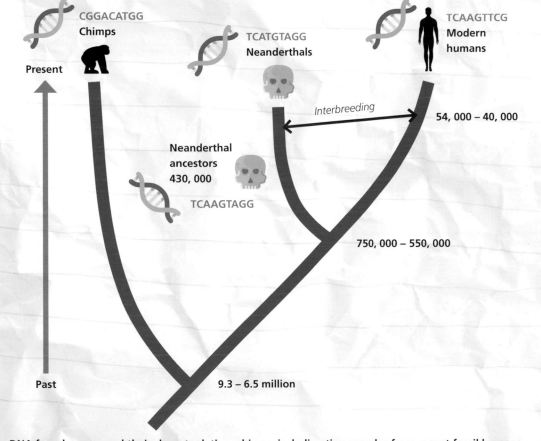

DNA from humans and their closest relatives chimps, including tiny samples from recent fossil humans, has been used to estimate divergence times on our part of the tree of life.

Life may have begun in the deep oceans around volcanic vents that were rich in many kinds of minerals. Today, unique life forms live around so-called "black smokers" on the ocean floor, including the tube worms around this vent in the Pacific Ocean.

The Origin of Life

Overwhelmingly, the evidence points to all known species having evolved from a single ancestor: the "last universal common ancestor", or LUCA. But what kind of life form was LUCA itself?

We may never know for sure. Even the oldest fossils, dating from over 3.5 billion years ago, look still too large and complex to be reasonable candidates. Experiments have demonstrated that the molecules of life, such as amino acids, can form from a mix of simpler chemicals likely to have been around on early Earth. Current theories suggest that deep-sea volcanic vents may have served as "hatcheries" for life by concentrating and energizing these mixtures. Life would have started to evolve by copying errors and natural selection when complex molecules could first replicate. But

there is a problem. The genetic system of today, at its heart, involves two kinds of complex molecules: DNA and protein. DNA is the replicator, and protein performs the tasks. However, proteins can only be encoded by DNA and, as we saw in chapter 5, proteins (in the form of catalytic enzymes) are needed for DNA to replicate. So which came first, if they both rely on one another?

Recent ideas suggest life could have started with neither. Rather, the first life was based on the intermediary between the two: RNA. Today, RNA is best known as the "go-between" messenger that is copied for DNA's genes to the place of protein manufacture (see chapter 4). But some forms of RNA have catalytic properties too. Many scientists now see the first age of reproductive life, billions of years ago, as an "RNA world".

Chapter 11
HOW WE CAN READ DNA

Testing for Genetic Diseases

The physical characteristics of a living thing can tell us a lot about the kinds of genes it carries, but tests are needed to be sure about the presence of specific genes. Genetic testing began with the search for ways to diagnose genetic diseases.

As we have seen, the effects of inherited genes on the developing body are powerful but also subtly complex. Genes interact in complicated ways, and it is often very difficult or even impossible to make clear-cut deductions about the kinds of genes carried by a body simply by observing its visible characteristics. Genes that have simple inherited effects, such as flower colour of pea plants or fur colour of mice, are in the minority. To explore the genetic makeup of individuals, a definitive test is needed that can tell us something about the specific genes present. These genetic tests emerged as scientists sought ways to understand and diagnose inherited diseases.

When William Bateson and others were reviving Mendel's theory of inheritance in the 1900s, one of his colleagues, a British physician called Archibald Garrod, was applying the growing field of genetics to the study of disease. In particular, he was interested in a disorder called alkaptonuria that was associated with an over-production of brown pigment that affected the colour of urine of his patients. Garrod noticed not only that symptoms in babies were appearing soon after birth, but also

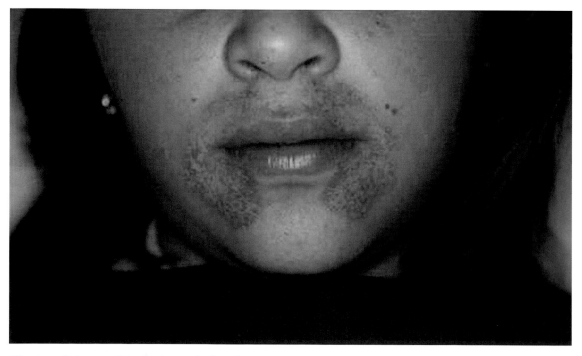

Alkaptonuria is a rare inherited genetic disorder. Symptoms can include urine turning brown, painful joints and pigmentation of the skin (above).

that the disease appeared to run in families. He knew that it was caused by abnormalities of chemical reactions in the body and he called such disorders "inborn errors of metabolism". Over the next half a century, the errors were attributed to gene mutations affecting proteins such as enzymes. Today, we know that alkaptonuria arises because of a mutation in a gene that ordinarily should encode for an enzyme that helps to process a certain kind of amino acid. In its absence, pigment-like chemicals build up in the body. But the fact that genes have specific chemical effects in the body meant they could, at least in theory, be detected by chemical tests.

Meanwhile, as experiments and observation down microscopes were revealing that the agents of inheritance, genes, were carried on threads called chromosomes, biologists were also beginning to recognize the significance of chromosome makeup in diagnosing other genetic disorders.

Looking at Chromosomes

The first unambiguous chromosomal studies were performed on organisms in which the chromosomes could be seen most clearly. These were the studies that demonstrated that a full set of chromosomes was needed for organisms to develop properly (see chapter 5). Once it was established that genes were carried on chromosomes, observations down a microscope became some of the first "genetic tests". But even by the 1950s, the makeup of human chromosomes was so difficult to resolve that no-one could even decide on the human chromosome number. Most people thought there were 48. A breakthrough came in 1956, when Indonesian-born geneticist Joe Hin Tjio discovered by careful microscope techniques that humans actually had 46 chromosomes.

Advances in resolving chromosomes and arranging their images into ordered sets, called karyotypes, allowed chromosomal abnormalities to be diagnosed. In 1959, French physicians Jerome Lejeune and Marthe Gautier discovered that Down's syndrome was caused by an extra chromosome: specifically an extra chromosome number 21.

Uncertainty about the human chromosome number persisted until the 1950s, when, in a eureka moment, better microscopy and staining techniques revealed that normal body cells contain 46.

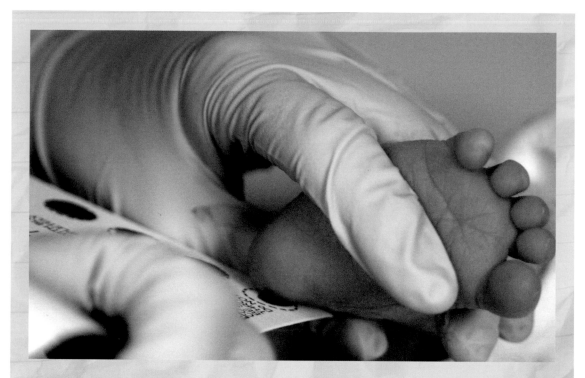

The first chemical test for a gene mutation called PKU was developed by Robert Guthrie. A blood sample taken from a newborn baby is spotted onto filter paper, which is then dried. Discs of the blood-soaked paper and bacteria are then added to a jelly for culturing the microbes. Visible colonies of bacteria grow on the discs if the blood contains excess phenylalanine, the indicator of PKU.

Chemical Tests for Gene Effects

Given the way that genes encode for proteins, some sort of chemical test could tell, at least in theory, if genes, or their mutated forms, are present in the body. For instance, a gene might encode for an enzyme. An enzyme is a kind of protein that works as a chemical catalyst: it drives a chemical reaction in the body that changes substance A into substance B. If the gene mutates and so fails to produce the enzyme, the reaction doesn't happen, so substance A would build up and substance B drop. Either could be detected by, say, a blood test or a measure of the deficiency in the enzyme itself. This would be the basis for testing the working of the gene.

In 1961, an American biologist called Robert Guthrie developed the first such test for an inherited metabolic defect. He focused on phenylketonuria (PKU). Like Archibald Garrod's alkaptonuria, PKU is a defect in the metabolism of amino acids, but its consequences are far more severe, leading to potentially life-altering symptoms such as seizures and heart defects. Specifically, it is associated with an accumulation of an amino acid called phenylalanine. Caught early, the disease can be controlled by a strict diet in which the unprocessed amino acid is avoided. But Guthrie needed a test that could pick up the accumulated phenylalanine in newborn babies. He developed one that used bacteria. A spot of PKU-blood added to the bacteria would make them grow into a visible colony. This property was developed into a simple tool that could be used as a routine test in natal units. The Guthrie test is still used today to screen newborns for PKU so that the dietary treatment can be prescribed.

How We Can Read DNA

Prenatal Screening

A variety of diagnostic tests are routinely used to detect genetic disorders in unborn babies or in newborns. Some have been developed specifically to screen for such diseases, while others make use of technology that has had a wider application in medicine. For instance, ultrasound uses high-frequency sound waves to produce an image of the body's internal structures. It relies on the fact that these waves bounce off organs and tissues in different ways. Ulstrasound is used to produce a scan of an unborn foetus so that physical abnormalities in development can be identified.

More invasive techniques are used to detect specific genetic disorders, through sampling of foetal cells, usually under the guidance of ultrasound. As a foetus grows, it is nourished by a placenta that develops in the wall of the womb. This controls the exchange of nutrients and waste between foetal and maternal blood, while preventing the two bloodstreams from mixing. Since 1983, physicians have been able to take samples of foetal cells from the placenta, a technique called chorionic villus sampling (CVS). Alternatively, foetal cells can be extracted from the amniotic sac that surrounds and cushions a developing foetus. This is called amniocentesis and it was developed in the 1960s. Although they are the most reliable methods of diagnoses, both CVS and amniocentesis carry a tiny risk of miscarriage. More recently, techniques have been invented that bypass the womb altogether and instead rely on taking a blood sample from the pregnant mother. It had long been known that, despite the placental barrier, a few foetal cells leaked into the mother's bloodstream. In the 1990s, these cells were isolated sufficiently well to detect foetal chromosome abnormalities.

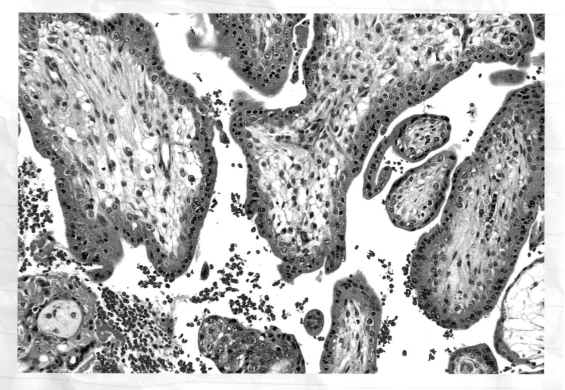

Foetal cells isolated from prenatal tests, including amniocentesis and chorionic villus sampling (as seen above), can be examined for chromosomal mutations, such as Down's syndrome.

Sequencing DNA

The earliest genetic tests relied on detecting the chemical products of genes or looking at chromosomes, but more sophisticated techniques were developed to test the DNA itself.

General chemical tests that could reveal DNA have been developed since the 19th century. Stains added to cells stick to DNA or chromosomes, and show them up as a distinct colour. These techniques are still used today, and are useful for showing that genetic material is found in the cell's nucleus, for instance, but they cannot be used to test for specific genes.

The biggest hurdle for directly testing genes is that genes are not physically separate things. Rather, a gene is a stretch of a DNA double helix that spans a length of bases on a much longer DNA molecule. Thousands of genes can be linked together on a single chromosome, and it is difficult to see how a simple chemical test could discriminate where one gene ends and another starts, let alone be specific for specific genes. Testing for genes directly is a world away from testing for a gene product in blood. Instead, scientists began to approach the problem of DNA testing by finding a way to analyse the sequence of its building blocks. If they could sequence the bases along DNA, that would be one step closer to finding a way of testing for genes. Fortunately, a British biochemist was already one step ahead, in the sequencing idea.

Sequencing Long Molecules

In 1952, Frederick Sanger was the first scientist to work out the sequence of building blocks along a long biological molecule: insulin. Although this was not DNA, the principles involved would be similar. Insulin, like all proteins, is made up of a chain of amino acids, and Sanger was the first to show that the sequence of amino acids of a particular kind of protein was very specific. He treated the insulin with chemicals that broke the chains up into smaller pieces, but managed to do it by stripping one amino acid off at a time. He could then identify the amino acids, one by one. He combined this technique with other methods that involved breaking the protein chain up into pieces and then looking for regions of overlap. Finally, the building blocks of the protein – either individual amino acids or short section of the chain – could be separated and identified. Sanger used a technique called chromatography for this. In the end, he worked out the sequence of amino acids in the entire insulin molecule.

British biochemist Frederick Sanger won two Nobel Prizes for Chemistry.

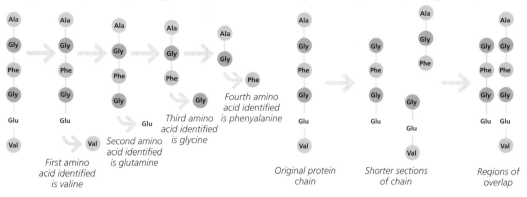

Sanger's technique for sequencing insulin part 1

First amino acid identified is valine

Second amino acid identified is glutamine

Third amino acid identified is glycine

Fourth amino acid identified is phenylalanine

Sanger's technique for sequencing insulin part 2

Original protein chain

Shorter sections of chain

Regions of overlap

One of the analytical techniques used by Frederick Sanger in working out the sequence of amino acids in the protein insulin involved snipping off its amino acid building blocks one by one and identifying them at each stage.

Sanger also broke the samples of the protein into multiple smaller overlapping lengths. By comparing the regions of overlap in the amino acids, this was able to help him decipher the sequence of the entire chain.

Paper chromatography of amino acids

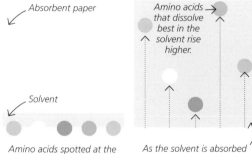

Absorbent paper

Solvent

Amino acids that dissolve best in the solvent rise higher.

Amino acids spotted at the bottom of absorbent paper, which soaked in solvent.

As the solvent is absorbed and rises through the paper, it takes the amino acids with it.

Techniques called paper chromatography, and modifications of them, are used to analyse the building blocks of complex molecules such as proteins. A mixture of amino acids (or short chunks of protein) are dabbed onto the bottom of absorbent paper; the bottom is then soaked in a solvent. As the solvent gets absorbed by the paper, it rises, taking the amino acids with it. Different kinds of amino acids, with unique chemical properties, rise to different heights: the ones that dissolve best in the solvent rise highest. Their spots can then be revealed by staining.

Applying Techniques to DNA

When Sanger turned his attention to DNA, he succeeded in developing a technique that helped him to work out the sequence of bases in the DNA of a virus that infected bacteria. This time, there was no way he could take the DNA apart, base by base. Instead, he came up with a way of working out the sequence in reverse: by building it up. Sanger made use of DNA's replication property, which relied on the DNA-replicating enzyme (called DNA polymerase: see chapter 5), which could be made to work on isolated DNA in the laboratory.

First, he replicated the DNA to obtain four identical batches. Then he mixed each batch with one of the four DNA bases (A, T, G or C), but used special modified bases that blocked the replication in its tracks. By seeing how far each batch was copied, he could work out the order of the bases in the DNA chunk. Today, computerized sequencing machines use Sanger's method to determine the base sequence of a DNA sample automatically, speeding up the analysis and minimizing the cost.

Sanger's virus DNA was 5,386 bases long across nine genes. Completed in 1979, this was the first time that anyone had successfully analysed the complete genetic makeup of anything. It meant that biologists could potentially understand how the virus worked from the basis of its genetic building blocks. But viruses are tiny, and the prospect of doing the same for something as complex as a human, with 3.3 billion base pairs and more than 20,000 genes, seemed overwhelmingly daunting.

Tracking Genes

The complete genetic makeup of an organism, cell or virus is called its genome. A genome consists of the entire sequence of bases, covering all the genes and the non-coding regions of DNA in between. Genome should not be confused with genotype (see chapter 6). A genotype is just the simple combination of one or a few allele pairs, such as tall pea plants being TT or Tt. A genome is the entire chemical makeup of all the DNA.

The entire human genome appeared to be an impossible target, scientists initially used others ways to make inroads into its secrets. One technique, called linkage analysis, relied on the fact that genes were linked together in sections on chromosomes, meaning that they tended to be inherited together. If you could identify a section of DNA that ran through a family with a history of a genetic disorder, there was a good chance that the gene for that disease was somewhere in that section, linked to adjacent areas of DNA that were inherited along with it. In 1983, linkage analysis successfully located the position of the gene for Huntington's disease, a disease of the brain (see chapter 3), to a region of chromosome 4. It was

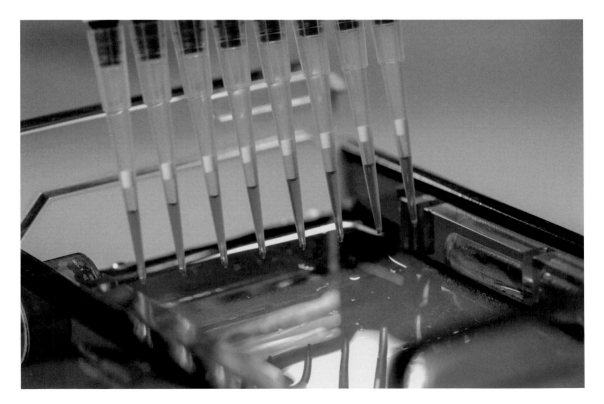

Biochemists use a technique called gel electrophoresis to separate sections of DNA or RNA by length. The molecules are separated by applying an electric field to them. Shorter molecules move faster and further than longer molecules. This technique is performed prior to DNA sequencing.

How We Can Read DNA

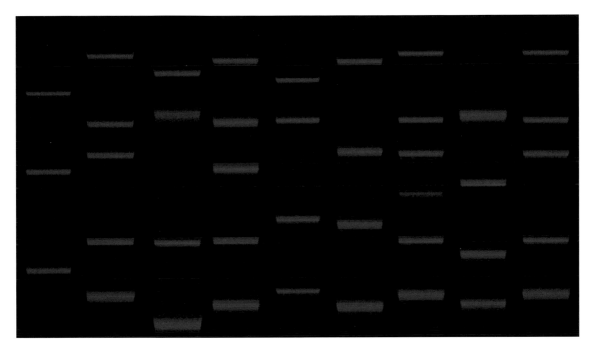

Genetic profiling (genetic "fingerprinting") relies on the fact that different individuals have variable sections of DNA called tandem repeats. Gel electrophoresis separates these sections to give a sequence of bands, like a bar-code. If two samples are exact matches, they must come from the same individual or from identical twins.

the first mutated human gene to be traced to a specific position on the human genome.

Genetic Fingerprinting

While the push for ways to sequence DNA and locate genes was making progress, other scientists were approaching DNA analysis with a different purpose in mind. In particular, they were looking for ways to compare the genetic information in DNA from different samples. There were many applications, from establishing parentage and other family relationships to understanding the genetic makeup of populations of animals and plants. But it was in the field of forensic analysis that one technique became especially well-known. It is called DNA profiling, popularly known as genetic fingerprinting.

DNA profiling was devised by British geneticist Alec Jeffreys in 1984. Jeffreys was focusing on variable regions of DNA called tandem repeats, which are stretches of repetitive base sequences –

like stutters in DNA's message. The sections are found in all individuals in a population, but they vary by the number of repeats: some people carry more repeats than others so their sections are longer. These sections do not usually correspond with genes, and only represent a tiny fraction of the genome, but genes are irrelevant if your only aim is to compare DNA from different samples. In forensic analysis, for instance, you might want to compare DNA from a murder weapon with the DNA in several suspects to find an exact match.

DNA from each sample is first broken up into fragments that correspond to the tandem repeats. Fragments with more repeats will be longer than others. A technique called gel electrophoresis is then used to separate the fragments by size, which are then marked with radioactive chemicals that will show up on an X-ray film. Each sample will produce a particular pattern of scattered marks, rather like a bar-code. The geneticist can then compare such profiles from different samples to find a match.

Cracking Genomes

With the dawn of a new millennium, the ultimate goal in genetic sequencing, cracking the entire human genome, became a reality. By 2003, an international collaboration called the Human Genome Project had produced a sequence that was 99.99 per cent accurate.

In the 1980s, when scientists started discussing the possibility of sequencing the entire human genome, nobody was under any illusions about the scale of the project. The previous decades had seen considerable advances in technologies that had

The Human Genome Project was a brainchild of James Watson, co-discoverer of the double helix structure of DNA, but was later headed by Francis Collins (above) who was in charge of the project from 1993 to 2008.

made the venture more realistic. A procedure invented by Kary Mullis, called the polymerase chain reaction (see chapter 5), could amplify tiny quantities of DNA by replicating it artificially in the laboratory. This could provide the huge amounts of human DNA that would be needed. And then there were the sequencing techniques pioneered by Frederick Sanger. Although the procedure was laborious, semi-automatic machines had been invented to speed it up. Nevertheless, the sheer scale of the venture, estimated to cost US$3 billion dollars, was huge. It was reckoned it would take 1,000 technicians 50 years to complete. The project was to become one of the single biggest collaborative efforts in the history of science.

The Human Genome Project, as it became known, would be the most ambitious aim for the science of genetics, but other projects to sequence the fruit fly and a microscopic worm would run concurrently. These smaller genomes would serve to support the human project, helping to guide the practicalities of their more complex counterpart. In 1989, the US National Institute of Health (NIH) was made the lead agency in the Human Genome Project, which was initially headed by James Watson, who had co-discovered the double helix nature of DNA back in 1953, but later directed by American geneticist Francis Collins. The work, involving the collaboration of scientific organizations around the world, began in 1990 with a projected timescale of 15 years.

Faster Sequencing

A biologist at the NIH called Craig Venter eventually became frustrated that the proposed methods of the Human Genome Project were so laborious. He favoured a quicker way forward that concentrated on the protein-encoding parts of DNA. Unlike the official Project approach, Venter wanted to ignore stretches of DNA between genes, as well as the non-coding portions of genes (called introns: see chapter 2). He also wanted to rely on a technique that randomly broke the DNA up into tinier fragments, which could then each be sequenced by more traditional methods. By repeating this many times over, a computer could then examine any overlaps and attempt to assemble the original sequence. The method, first pioneered in the early 1980s, was called shotgun sequencing.

In 1992, Venter left the NIH to establish his own private genome research company. By 1995, he had succeeded in using the shotgun technique to sequence the genome of an influenza bacterium. This was the biggest genome yet sequenced. At more than 1.8 millions base pairs and 1,740 genes, it was considerably larger than Frederick Sanger's previous breakthrough sequencing the genome of a virus.

Biologist Craig Venter (above) left the NIH to set up a privately funded rival company.

Modern sequencing methods, called Next Generation Sequencing, mean that scientists can now read DNA far more quickly than even just a few years ago.

Multi-Celled Genome Projects

The Human Genome Project adopted a more orderly approach than Venter to the mammoth task that lay ahead. The scientists at NIH were concerned that Venter's more random shotgun methods would leave gaps in the sequence. In the end, both parties embarked on the gargantuan task, each with their different competing strategies.

By 1998, another breakthrough came, this time by the official Genome Project group. They had sequenced the first multi-celled organism, a microscopic roundworm called *Caenorhabditis elegans*: the "elegant rod-worm". Scarcely a millimetre in length, this tiny worm lives in compost and rotting matter, where it eats bacteria. Most are hermaphrodites; a few are males. Like the modest fruit fly and garden pea, it has an iconic

place in the history of biological research. Unlike bigger organisms such as humans, the body of the rod-worm always has a fixed number of cells: 959 in the hermaphrodite majority; 1,031 in the minority males. Painstaking studies in the 1970s managed to trace the developmental pathway of each cell. With the publication of its genome, the rod-worm became, arguably, the best known multi-celled organism in existence.

The rod-worm genome had just over 18,000 genes, slightly more than a third of which were similar to genes in humans. And the vast majority of those similar genes – about 90 per cent - were somehow involved in keeping the multi-celled body together. A year later, Craig Venter's organization announced that it had completed sequencing a second multi-celled organism with 13,601 genes: the fruit fly.

The tiny Elegant Rod-Worm (*Caenorhabditis elegans*) was chosen as the first target organism for sequencing a multi-celled species. With the completion of its genome project, plus a "map" of all its body cells published in previous decades, it became one of the best known of all living things.

How We Can Read DNA

In 2012, scientists at the University of Leicester, UK, created a printed version of the human genome, containing some three billion letters and printed with 43,000 characters per page. It filled some 130 volumes.

The Human Genome

On 26 June 2000, in a presentation at the White House, Francis Collins and Craig Venter jointly announced their "first survey" or "rough draft" of the human genome. There was still a lot to be done to fill the gaps, but much of the human base sequence was finished. The scientific papers reporting the details of their achievements were published in February the following year. It would take another two years before the entire human genome was completed in April 2003, but that was still two years earlier than expected. But now that the map of human genetic information was known, considerable work remained on continuing to decipher what the information meant and how the many genes controlled the human body.

The publication of the Human Genome Project was a landmark in the history of human genetics. We now know more about the genetic makeup of our species and in far greater detail than ever before. Human DNA is arranged into 46 chromosomes, arranged into 23 pairs. This DNA spans 20,687 genes (according to a current source). This alone was an astonishing

discovery, far fewer than previously suspected. Many kinds of organism have many more genes than humans. Clearly, the key to understanding the nature of living complexity lies in the kinds of genes, rather than numbers.

The first gene on chromosome number 1 determines a protein needed to smell. The last gene, on chromosome X, helps to control the immune system. In between, genes with very different jobs seem to be randomly placed, but some, such as those involved in embryonic development, are clustered together.

Human DNA spans about 3.2 billion bases running along one side of all the DNA double helices. Remarkably, 98 per cent of human DNA is not gene. In other words, the vast majority of DNA is made up of long stretches of DNA between genes that encode for protein, or as the "nonsense" patches called introns that exist at intervals inside genes (see chapter 2). Many sections of the human genome, including entire stretches 300 bases long, are repeated over and over again. The significance of these repeats is not yet known.

Using information gathered from the Human Genome Project, scientists hope to identify the genetic causes of certain aspects of human behaviour, including those linked to excitement or thrill-seeking.

Genetic Identity

The Human Genome Project has opened up the possibility of answering some of the oldest questions in human genetics. The genome map has enabled scientists to pinpoint genes involved in disease, and the more that is known about these genes, the more likely it is that treatments or even cures will be found. But does the genome have anything to say about more complex aspects of human behaviour, such as temperament, sexuality or mental illness?

A wealth of evidence is needed to understand how behaviour is controlled by genes. That a degree of genetic control is involved is indisputable, because the nervous system, including the brain, develops under the influence of controlling genes, like all other parts of the body,. The issue of genetic effects begins by examining family histories. If these show a pattern of inheritance for, say, mental illness, then this suggests genetic control. But inheritance for complex traits, as we have seen, rarely follows the simple pattern that was uncovered by Gregor Mendel a century and a half ago. Traits shared by siblings, and especially identical twins, are very likely to be determined by genes. Then it is necessary to turn to results of analysis of the DNA itself to see if there are any common factors lying among the sequences of bases. For instance, studies of male homosexuality initially implied an inheritance pattern that pointed to a "gay" gene in the X chromosome. But now it seems that multiple sections of DNA are involved and in a variety of locations. Undoubtedly some individual genes might have a great influence on complex human behaviour. For instance, scientists searching for a genetic component for temperament have focused on a gene that controls a kind dopamine receptor in the brain. This may affect how the receptor creates a "risk-taking" or "novelty-seeking" temperament. But, in reality, the real challenge that lies ahead is to use the map of human genetic information to understand better how genes interact in complex ways with other genes, and also with the environment around them.

Chapter 12
HOW WE CAN
MANIPULATE GENES

Artificial Selection

When humans breed animals or plants and select certain characteristics, we are mimicking what nature does in driving the evolution of new kinds of organisms. This artificial selection has produced domesticated breeds to improve the quality of human lives.

Humans have been manipulating the genetic makeup of other living things since before the dawn of written history. About 10,000 years ago, prehistoric *Homo sapiens* abandoned their hunter-gatherer lifestyle in favour of a more sedentary existence. As these people settled into fixed communities, they began to use the natural resources around them in a more intensive way. Perhaps, after seeing seeds sprout in middens, they decided to grow their own food plants or transplant food plants closer to home. At the same time, they corralled useful animals for meat and milk, as domestic companions, or as pack animals. They learned how to breed their charges and over time, whether consciously or unconsciously, they selected the most useful or beneficial individuals to breed from. Unwittingly, these prehistoric farmers became the first geneticists.

When humans selectively breed for particular characteristics in animals or plants, this is called artificial selection. When the resulting breeds became reliant on their human farmers, they are said to be domesticated. Today, it is tempting only to associate genetic manipulation with clinically precise procedures involving laboratories and test tubes. But it actually began with artificial selection thousands of years before the work of anyone who could be called a scientist.

Despite their obvious differences, all living breeds of domesticated dogs belong to the same species as the wolf: they have much the same genes and chromosomes. Over the last few hundred years, intensive programmes of selective breeding have exaggerated certain traits, such as body size, shape and temperament.

How We Can Manipulate Genes

Artificial selection has continued to change the characteristics of cultivated wheat to make it more useful to farmers. Seed heads were selected with more nutritious seeds that did not shatter during harvest, while plants were bred to be more resistant to disease.

The First Selectors

The first plants to be domesticated were members of the grass family, such as wheat, barley and rice. Originally, humans relied on wild populations of these grasses, eating their seed heads, which were rich in carbohydrates. But when these grasses were cultivated, the early farmers took advantage of the variation that existed in the wild populations: they selected the plants with the fattest seeds and those that were easiest to harvest. Over many generations, the plants came to acquire these characteristics. The grass plants ended up producing more nourishing seeds that were carried on stronger heads that did not shatter prematurely before harvest. These prehistoric geneticists were creating the first breeds of plants that would become staples in the diet of humans today.

Animals, too, were selected along similar lines. Dogs were probably domesticated more than five thousand years before the first crop plants. They were selected for their temperament, as they helped in hunting expeditions or defended campsites. Only the most docile animals would have bred alongside their human companions, something that their human carers exaggerated by artificial selection.

With a modern understanding of genetics, it is possible to see how these domestication pathways succeeded. The fact that many of these characteristics, including seed size and animal temperament, are influenced by genes means that artificial selection over thousands of years could make those selected populations diverge a long way from their ancestral wild populations. Today, cultivated wheat and wild wheat look very different. All modern breeds of dog are regarded as belonging to the same species, even though the difference between, say, a chihuahua and a wild wolf looks very great indeed.

New Breeds From Old

The oldest known production of a domesticated variety occurred in wheat plants. A wild species called einkorn wheat can still be found in the eastern Mediterranean region today, but 10,000 years ago, selective breeding by prehistoric farmers produced a domesticated variety that retained its seeds for longer. This was good for an efficient harvest, but not good in the wild: such seeds would not properly disperse. It meant that seed-retaining einkorn only thrived in cultivation.

The genetic history of modern wheat involves a story that is dominated by changes to chromosomes. The fundamental chromosome makeup of wheat plants and related species of grasses is a diploid number of 14 chromosomes: two sets of seven. About half a million years ago, a species emerged with double this number. Durum wheat had evolved by polyploidy (see chapter 10): it combined two diploid arrangements to give a tetraploid (four-set) arrangement of 28 chromosomes. This gave it extra-sticky protein, which made it a poor choice for making bread; instead, durum wheat today is used for making pasta. But the chromosomal multiplication did not stop there. Tetraploid durum wheat hybridized with a wild diploid grass to produce a wheat plant that was hexaploid: it contains six chromosome sets. This chromosome-packed grass is the common bread wheat that would become the most widely grown crop in the world. Today, 95 per cent of all the globe's wheat comes from this hexaploid.

Modern crop plants have been bred and developed to maximize their productivity, their resistance to pests and the ease with which they can be harvested.

How We Can Manipulate Genes

The properties of common bread wheat made it a perfect choice for baking, and this species was favoured by early farmers for that reason. It soon largely displaced wild forms such as einkorn around the globe. It contained just the right amount of protein to make the dough elastic but malleable, to form a structure that helped it to rise. Humans continued to select the most high-yielding wheat plants that were more manageable in cultivation, and artificial selection has continued apace in the modern age of genetics. In the 1960s, selective breeding experiments undertaken by American agronomist Norman Borlaug produced strains of wheat that were resistant to disease and had shorter, thicker stems. This meant that their larger seed heads did not collapse under cultivation. Borlaug's wheats went on to be grown as staple crops throughout the world, and he achieved similar successes in improving rice. In 1970 he was awarded the Nobel Peace Prize for his contributions to the global food supply.

The Principles of Selective Breeding

Selective breeding involves choosing individuals with the best characteristics according to your requirements, and using these to produce offspring. Good traits tend to occur in related individuals where the genes for those traits are more common. If the best characteristics are determined by recessive forms of genes, they will only be expressed when two of these identical forms, or alleles, come together in a so-called homozygous combination. This practice of interbreeding genetically related individuals is called inbreeding, and is necessary to establish new target traits in a population.

However, inbreeding also causes problems. In addition to the desirable traits, other less desirable ones, also determined by recessive versions of genes, may also be expressed. Inbreeding, when carried to extremes, can make a population homozygous for lots of recessive genes, including those that cause disease or disorders. As a result, the health and well-being of the population declines, something called inbreeding depression.

Plant and animal breeders minimize the damaging effects of inbreeding by culling the weaker individuals and using occasional cross-breeding, with other desirable strains. Cross-breeding helps to introduce new alleles into a population that will mask the effects of harmful ones. Hybrid lines of this sort are said to carry "hybrid vigour": technically, they are said to have heterosis. In practice, then, a careful combination of in-breeding and out-breeding are necessary to produce strong, useful strains of plants and animals.

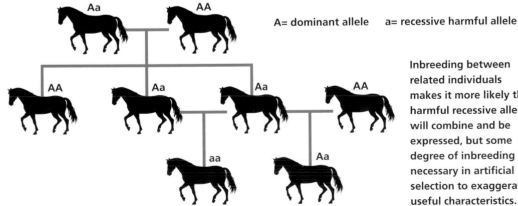

A= dominant allele   a= recessive harmful allele

Inbreeding between related individuals makes it more likely that harmful recessive alleles will combine and be expressed, but some degree of inbreeding is necessary in artificial selection to exaggerate useful characteristics.

Manipulating Microbes

Knowledge of the molecular systems involved in copying and expressed genes is now so sophisticated that scientists can go far beyond selective breeding in modifying the genetic makeup of cells and organisms. They started with organisms with the simplest genetic systems: microbes.

In the long term, artificial selection is the easiest way for humans to change the genetic makeup of living things. By controlling the way plants and animals breed, we can produce plants capable of feeding humanity from wild grasses, produce prize-winning dogs and thoroughbred horses. But the process can be laborious. Whole plants and animals are complicated mixtures of thousands of genes, and getting a gene combination that is right takes time and is never perfect.

In the 1970s, geneticists began to think about more invasive ways of manipulating genetic makeup. Within just a couple of decades of discovering the double helix structure of DNA, the progress of molecular biology had been staggering. Scientists had deciphered the genetic code for life, worked out how genes could replicate, and understood how their information was translated into protein and, ultimately, characteristics. Could genes be manipulated in the laboratory just like any

other kind of chemical substance? Genes were made of DNA, and DNA, together with the catalytic enzymes on which it depended, behaved according to the ordinary rules of chemistry. Geneticists thought that it might be possible to control these chemical reactions and move genes from one cell to another.

In 1971, the US National Institutes of Health (NIH) organized a conference under the title "Prospects for Designed Genetic Change", which served to focus attention on a branch of genetics that was, at least then, more science fiction than science fact. But biologists were excited by the possibilities. Characteristics of organisms could be changed in ways that were more precise than by using selective breeding; they would engineer the genes directly. And, in the world of medicine, it might even be possible to cure genetic diseases. The conference ushered in an age of genetic engineering.

Even though DNA is an amazingly complicated piece of kit, the toolkit that scientists use to manipulate it can be surprisingly simple.

How We Can Manipulate Genes

The Tools of the Trade

DNA, and indeed the entire molecular machinery inside living cells, relies on chemical catalysts called enzymes to engage in reactions. Enzymes are protein molecules with particular shapes that lock onto their targets and help to drive then towards a chemical change (see chapter 3). Each kind of enzyme only locks onto a target with a particular complementary shape, meaning that enzymes, in all their countless different forms, are highly specific in the processes they catalyse.

After the discovery of the double helix structure, scientists had worked hard to trace the enzymes involved in replicating and repairing DNA. These and related enzymes would become part of the toolkit of genetic engineering. Three kinds of enzyme were especially important. First, there was DNA polymerase that helped to build new DNA whenever it replicated. Polymerase brought together nucleotides, the building blocks of DNA, bonding them into long chains as one double helix became two (see chapter 5). Second, there was DNA ligase. This was an enzyme that sealed gaps in DNA, used in completing the replication process and to help repair damaged DNA. These enzymes would be useful in fixing genes in place, but how would they be extracted?

The third kind of enzyme in the toolkit came from a very particular place. Polymerase and ligase are found in all living cells, but enzymes that cut DNA in very predictable ways are mainly found bacteria. These enzymes evolved in bacteria as a defence armament against other microbes, specifically viruses that could infect their cells. Their DNA-cutting enzymes helped to target specific viral DNA that was identified as foreign. They had been called restriction enzymes because they restricted the activity of the invading virus.

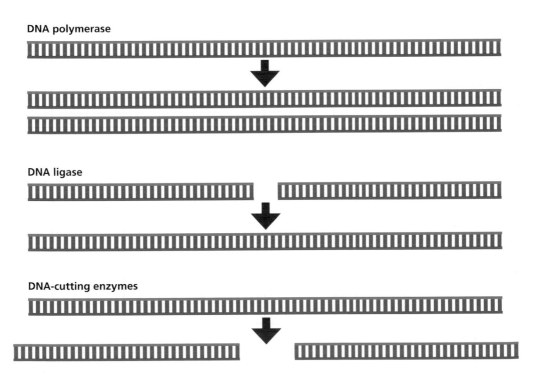

The basic genetic engineering toolkit consists of three kinds of enzymes: one to build DNA, one to seal gaps, and one to break it open.

The First Genetically Engineered DNA

The first attempts at engineering DNA using the enzyme toolkit were experimental. In 1971, American biochemist Paul Berg was the first scientist to join DNA together from different kinds of organisms: from viruses and bacteria. He called the hybrid he had created "recombinant DNA". Natural recombination of genes by sexual reproduction was a routine part of life, but this was the first time that genes had been chemically recombined by artificial means.

Two years later, two more American biochemists, Herbert Boyer and Stanley Cohen, took the next step: they used the same DNA-manipulating technology to produce the first genetically engineered cells. Because of the increasing rise of safety concerns, they confined their efforts to a process that mimicked something that happened naturally in the world of reproducing microbes. Bacteria routinely exchange rings of DNA called plasmids (see chapter 5). These rings carry genes that are useful for the bacteria. For instance, some make them resistant to chemicals such as antibiotics. Boyer and Cohen used the enzyme toolkit to cut one such resistant gene from one plasmid and insert it into another. Restriction enzymes were used for the cutting and ligase was used for stitching the DNA back together.

The resulting hybrid plasmids were mixed with bacteria, which absorbed them. The bacteria became antibiotic-resistant, just as though they had gained the genes naturally. Boyer and Cohen had demonstrated that the enzyme toolkit really could be used to genetically engineer living cells.

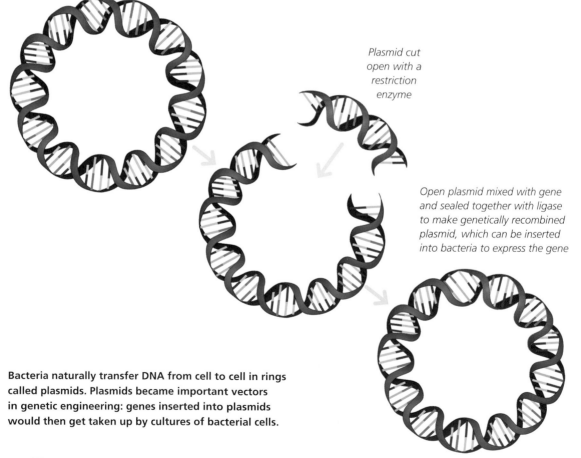

Plasmid cut open with a restriction enzyme

Open plasmid mixed with gene and sealed together with ligase to make genetically recombined plasmid, which can be inserted into bacteria to express the gene

Bacteria naturally transfer DNA from cell to cell in rings called plasmids. Plasmids became important vectors in genetic engineering: genes inserted into plasmids would then get taken up by cultures of bacterial cells.

How We Can Manipulate Genes

Useful Products

Although the long-term possibilities of genetic engineering were diverse, the initial focus was on something that seemed achievable within the confines of the fledgling technology: the production of useful protein. Proteins are naturally expressed by the genes inside living cells and perform all manner of tasks. Many are enzymes, while others are hormones: the chemical messages that circulate in the bloodstream. Some of these hormones, such as insulin, are of medical importance in treating diseases, including diabetes. But medicinal insulin had to be extracted from the pancreases of cows and pigs, a very inefficient production line that could also carry the risk of disease. Insulin is encoded by a gene found only in animals, but if the gene could be inserted into bacteria, the microbes could take over its commercial production.

A large vat of bacteria could, in theory, churn out enough insulin to satisfy the demand from diabetics. Herbert Boyer set up a company to do just that, with the intention to make the necessary genes that could be inserted into bacteria. But insulin is a complex protein made up of two different chains joined together, containing 51 amino acid building blocks in total. This meant it was controlled by two hefty-sized genes. Boyer needed something simpler to start with, so he began with a smaller growth-hormone protein, just 14 amino acids long, called somatostatin. As with insulin, its amino acid sequence was known from sequencing techniques that had been developed earlier (see chapter 11). This meant that scientists could use the genetic code to work out the base sequence of the gene that encoded for it. Boyer's lab then assembled DNA with this base sequence using the appropriate nucleotides: DNA's building blocks. The assembled gene was inserted into a plasmid, which was taken up by bacteria. When the bacteria reproduced, the somatostatin gene was replicated with them. By 1977, the team had succeeded: their bacterial cells were producing recombinant somatostatin. A year later, they had similar success with the bigger challenge that was insulin. Bacteria produced the two insulin chains separately, and chemists joined them together to make a working hormone. Recombinant insulin had arrived, and went on to be used commercially. Today, it is used to treat diabetics all around the world.

Using genetically modified plasmids to produce useful protein

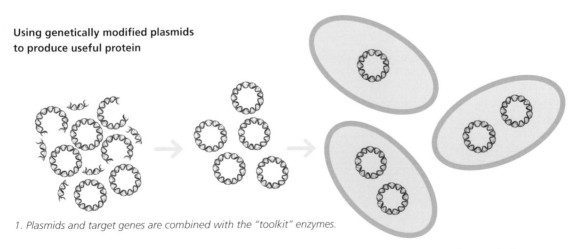

1. Plasmids and target genes are combined with the "toolkit" enzymes.

Genetically modified plasmids are easily taken up by bacteria, and can be used as a way of getting target genes inside the microbes. Once there, the gene is expressed by the bacterial cells to produce the protein encoded by the foreign genes.

2. This creates the genetically modified bacteria.

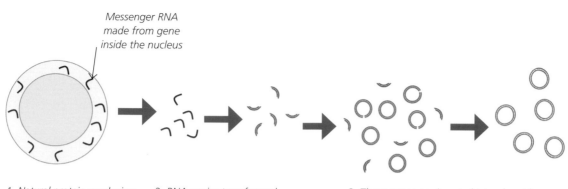

Messenger RNA made from gene inside the nucleus

1. Natural protein-producing cell producing copies of the messenger RNA

2. RNA copies transformed into complementary strands of DNA

3. These genes are inserted into plasmids, which are taken up by bacteria to make the protein in the usual way.

An enzyme called reverse transcriptase is used to produce a "slimmed-down", edited version of a large gene so that it can be inserted into bacteria, which cannot "read" the unedited genes found in human cells.

Targeting Bigger Genes

Both somatostatin and insulin were produced using genes that had been artificially assembled in the lab, base by base. But many genes were much longer, and building them up from scratch in this way was not practicable. In the 1980s, this became a big problem when Boyer's lab turned to their next protein target: factor VIII, the protein that helps to make blood clot properly. At the time, haemophiliacs were being treated with factor VIII that had been extracted from donated human blood. But the possibility of disease transmission became a reality in the 1980s, as many stocks were contaminated with the HIV virus. As a result, many haemophiliacs developed AIDS.

The protein factor VIII is especially large: it contains 2,350 amino acids, nearly 50 times more than insulin. This meant that building its gene from scratch was simply unworkable. So could scientists somehow extract the factor VIII gene ready-made from human cells? Unfortunately, there was a problem here, too. Like all genes from complex cells, human genes contain patches of non-protein coding DNA called introns (see chapter 2). When human cells naturally make their proteins, these

introns are edited out when they make their RNA message from the DNA gene. But bacteria do not have introns in their DNA, so they lack the ability to edit. If the human factor VIII gene were inserted, unedited, into bacteria, the microbes would find the gene's message garbled. No useful protein would emerge.

The answer came from a different kind of microbe: a virus. Some kinds of viruses, called retroviruses, contain just RNA instead of DNA. But they carry their own special enzyme that can make a DNA copy of their RNA when they infect host cells (see chapter 5). This enzyme, called reverse transcriptase, is used to reverse the RNA message back into a DNA gene. If Boyer's lab could extract the RNA messages for factor VIII from human cells, they could use the enzyme to reverse the messages back into lots of DNA genes. It worked, and in 1983, they produced plasmids engineered with human factor VIII genes, paving the way for the commercial release of a safer, disease-free blood-clotting agent. Meanwhile, the virus enzyme reverse transcriptase was added to the genetic engineering toolkit.

How We Can Manipulate Genes

Manipulating Plants and Animals

The cells of plants and animals are more complex than bacteria in many different ways. This makes their genetic engineering more challenging, but in the fast-paced field of genetics, scientists found ways to do it.

Plants and animals are eukaryotes. This word comes from the Greek: *eu*, meaning "true" and *karyon*, meaning "kernel". The DNA of each of their cells is packaged into a membrane-bound sac called a nucleus. Bacteria are prokaryotes (*pro* meaning "before") and lack a nucleus. Eukaryotes are also packed with much more DNA, with many times more genes than bacteria. Many of the extra genes are needed to control the assembly of a complex multi-cellular body. Eukaryotic DNA is additionally supported by molecules called histones. These are the proteins that coil the DNA so that chromosomes appear during cell division. The genes of eukaryotes are also peppered with non-coding "nonsense" patches called introns, which must be edited out before proteins are made. Neither histones nor introns are found in bacteria. All these differences created obstacles when genetic engineers turned from bacteria to plants and animals.

Genetic Modification By Grafting

For all the explosive growth of techniques involving molecular biology that have been developed since the 1950s, the most recent research in plant genetics suggests that humans had been causing genetic engineering for thousands of years. Grafting was developed by some of the earliest farmers as a way of fusing useful characteristics on one plant without the labour of selective breeding. For instance, the shoot of a variety of tree that produced tasty fruit could be grafted onto another with disease-resistant roots. The resulting chimera would carry the best features of both.

Today's technology has revealed that, once a graft "takes", the cells from each variety of plants can exchange their nuclei or energy-processing structures such as mitochondria and chloroplasts. Mitochondria and chloroplasts contain tiny amounts of genes. This means that the plant partner varieties each contain hybrid mixtures of genes.

Deliberate artificial genetic modification of eukaryotes became a reality in the 1920s, when American geneticist Hermann Muller induced mutations in fruit flies by exposed them to X-rays. Later, scientists would discover that this kind of treatment can change the genes in all kinds of organisms. But this was a big step away from moving genes from organism to organism. Also, the genetic modification achieved through grafting or X-rays was random. Scientists had no control over specific genes.

Grafting is a way of combining useful characteristics of different kinds of plants, and recent molecular research has shown that cells of each partner end up exchanging genes.

The First Transgenic Animal

In 1974, the first genetically engineered animal was produced by German-born biologist Rudolf Jaenisch. It came about as a by-product of his research into viruses and, like in the early days of microbe engineering, was achieved by mimicking something that happened routinely in nature. Retroviruses are natural genetic engineers: after infecting cells, they incorporate their own genes into the chromosomes of the host cells (see chapter 5). Jaenisch was studying a retrovirus that caused cancerous tumours and wanted to know why infected mice developed tumours in some parts of their body, such as bones and muscles, but not others, such as the liver. Perhaps the virus just did

not attack liver cells. To test the idea, he needed a way of ensuring that all the cells of a mouse got infected. He did this by injecting the virus into early mouse embryos. As they grew, all the body cells developing from the embryos would be affected. In fact, his research showed that the virus genes were shut down during development, so none of the mice developed tumours. But the experiment produced the first animals born that had been completely genetically modified by artificial means. Later, they would be called "transgenic" animals.

Manipulating Plants with Bacteria

Using viruses was a way of getting genes into cells that by-passed the technical constraints of cutting DNA and stitching it back together. The viruses were naturally evolved vehicles for transmitting genes into host cells, just as plasmids were the natural vehicles for transmitting genes between bacteria. This meant that both viruses and plasmids could be looked upon as vectors for carrying genes, a bit like the way organisms are vectors for carrying certain diseases.

The use of vectors took us a step closer to controlled engineering of eukaryotic genes with the discovery, in 1977, that some kinds of bacteria naturally transmit their DNA into plants cells. *Agrobacterium* is a microbe that causes crown gall disease in certain species of plants. It is a pest in crops such as fruit and sugar beet. When it infects plant cells, a piece of DNA from its plasmids is inserted into the plant's genome, and its effect is to cause the growth of tumours. Scientists saw that *Agrobacterium* could be used as a new kind of vector for genetically modifying plants. All they had to do was insert a desired gene into its plasmid, and the microbe would do the rest. The piece of DNA that was tumour-inducing could be disarmed, so that only the useful gene took hold.

In 1983, the first plant genetically engineered to contain a specific gene was produced. Using *Agrobacterium,* a gene for antibiotic resistance was inserted into tobacco plants. This gene was chosen as a way of checking whether this kind of genetic

Working alongside fellow scientist Beatrice Mintz, Rudolph Jaenisch (above) produced the first transgenic animal when they injected retrovirus DNA into mouse embryos.

How We Can Manipulate Genes

Agrobacterium is a microbe that causes crown gall disease (above) in certain species of plants. It is widely used today to insert genes into plants.

modification was feasible. Unmodified plants cells grown in the laboratory could be killed by exposure to the lethal antibiotic. But the cells treated with the modified *Agrobacterium* would survive because they would have the resistance. The test case worked.

Today, *Agrobacterium* is an important vector in the engineering toolkit and is the commonest way of producing genetically modified organisms. In 2000, it was used to genetically engineer rice to combat vitamin A deficiency, a condition that kills hundreds or thousands of children around the world each year, and leaves many others blind. The genetically modified rice, called golden rice because of its colour, contains the yellow-orange plant pigment beta-carotene, the pigment found in carrots. *Agrobacterium* plasmids were engineered with genes from daffodils and soil bacteria that, together, controlled the manufacture of the pigment, which is used to build vitamin A by humans when eaten.

The bacterium *Agrobacterium tumefaciens* (above) is widely used as a vector for genetically modifying plants. Its plasmid, called "Ti", can be engineered to incorporate foreign genes that are expressed in the plant.

Agrobacterium could only be used on plants that were naturally susceptible to its infection. In 1987, biologists at the Agricultural Experiment Station in Geneva devised a solution to this, which was, literally, ballistic. They invented a "gene gun" that fired gene-coated tungsten particles into plant cells, just like a shotgun, and the genes worked. The gene gun technique was indiscriminate: it offered a way of potentially getting genes into all sorts of plant cells, and perhaps animal cells too.

Using Stem Cells

A key issue in genetically modifying plants and animals is to ensure that any inserted genes end up in all the cells of the body. Ideally, this means starting with a single cell, modifying it by adding a gene, and then letting the cell divide and develop to form an entire body. Cells produced by growth are clones of one another, since they result

from DNA replication. This means that they are genetically identical, so all will contain the modifying gene. This is exactly what happens with the body of a plant or animal that grows naturally. Plants cells can be treated in the laboratory with chemicals called growth regulators. These stimulate the cells to develop all the parts of the adult body, such as leaves and roots. The early stages of the technique, called micropropagation, are performed under sterile conditions before the cloned plants can be potted into soil.

But animal bodies do not grow from any cells quite so easily. Only stem cells are capable of producing new tissues and new bodies. There are stem cells in an adult body, but they are specialized to produce only cells of a particular kind. For instance, stem cells in bone marrow produce blood cells. Apart from the original fertilized egg, the only stem cells that can produce all the parts of a body (technically called totipotent stem cells) are found in the early embryo. Embryonic stem cells are extraordinarily versatile. When an embryo is just a tiny ball of cells, the cells can be separated, and each can be grown into a cloned individual. Moreover, these stem cells can be cultured in the laboratory.

Rudolf Jaenisch's first transgenic mouse was created by modifying embryonic cells, but the virus technique he had used could insert the gene anywhere in the mouse genome, potentially disrupting the function of the cell's native genes. By 1981, biologists had devised ways of overcoming this problem by using purer sources of DNA. New techniques for culturing stem cells offered a way of cloning genetically modified mice. The genes inserted were passed on to subsequent generations, something that Jaenisch's earlier techniques could not easily have achieved. A decade later, a variety of methods were being used to create strains of genetically modified mice. Many of these strains were "knock-out" mice: their genomes had been modified by knocking out the function of an existing gene. It meant that they developed disorders such as cancer. This technology would later play an important role in Genome Projects to decipher the workings of genes that had been sequenced.

How We Can Manipulate Genes

Manipulating Humans

When scientists first started to think about the possibility of genetically modifying organisms, one particular goal had always been prominent. If human cells could be engineered, it would offer hope to millions of people suffering from diseases that were caused by defective genes.

Once biologists found that they could genetically manipulate mice embryos with such apparent ease, attention quickly turned to humans. Biologists attempted to do the same thing with human embryonic stem cells that they had done with ones from mice. But they came up against obstacles. Humans stem cells were not so easy to manipulate in culture.

At the same time, the ethical implications of genetic modification were beginning to dawn on a cautious world faced with these brash new technologies. Regulatory authorities, such as America's National Institutes of Health (NIH), took a hard line when it came to modifying genes in humans. Nevertheless, in 1980 an American biologist called Martin Cline became the first

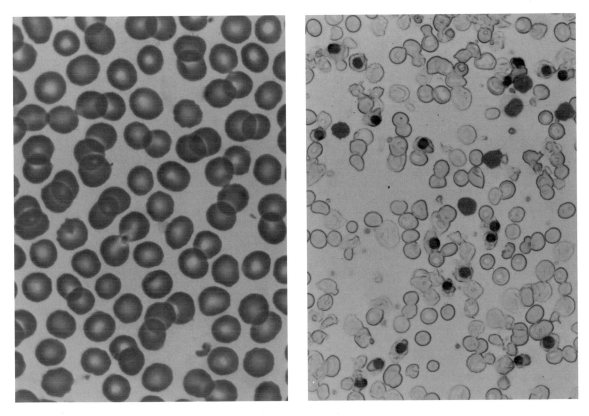

These micrographs show normal red blood cells (left) and red blood cells affected by the genetic conditions beta-thalassaemia (right). The condition reduces the production of haemoglobin, the iron-rich protein in red blood cells that carries oxygen to the body's cells.

researcher to attempt it, but he did it in Israel and Italy, away from the reach of the NIH. Cline was targeting a blood disorder called beta-thalassaemia, which causes severe problems in the liver and heart and is particularly common in the Mediterranean area. He successfully inserted DNA into the bone marrow of beta-thalassaemia sufferers, but the results of his trial were never published. When the NIH discovered what he had done, Cline was forced to resign from his position at the University of California in Los Angeles. Nevertheless, this marked the beginning of a new kind of genetic modification, one that would use it to treat human disease. The age of gene therapy had arrived.

Modifying Body Cells

Work on producing transgenic mice had always targeted embryos. That way, the inserted genes would be incorporated all over the body as the body grew and developed. And by beginning with isolated stem cells, these would grow into mice where every cells of the body was modified. The inserted genes would then become part of the mouse's germline: the genes are present in sex organs, so pass into sperm and eggs and are inherited, generation after generation. But ethical considerations, as well as practical issues involving "difficult" stem cells, prevented any such work on humans.

Researchers, therefore, looked towards the next-best thing in gene therapy. Instead of changing the fundamental germline, perhaps they could target specific tissues or organs in the body instead? It was this route that had been attempted by Martin Cline in his unofficial trials. This is called somatic, rather than germline gene therapy. It would not be as persistent, because cells in the developed body eventually weaken and die, so the gene therapy would have to be regularly boosted with a "top-up" of genes. But the therapeutic effects could, nevertheless, be effective. The first official trial was completed in 1990 by American biologists William Anderson and Michael Blaese, working on a disease called adenosine deaminase (ADA) deficiency. The defective gene means that suffers cannot make an enzyme (deaminase) to convert adenosine. As adenosine builds up in the body, it poisons the defensive white blood cells of the immune system. The immunodeficient children rarely survived into adulthood.

Genetically modifying the developing cells inside bone marrow could lead to the development of a wide range of blood cells.

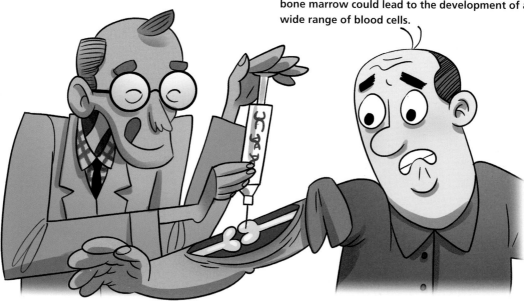

How We Can Manipulate Genes

The first approval trial of gene therapy was to treat a child suffering from an immunodeficient disorder caused by a mutation in a key enzyme-producing gene. The trial involved administering a working gene through a blood transfusion.

Initially, Anderson and Blaese wanted to use a technique that combined the use of retrovirus vectors and stem cells. Their idea was to extract stem cells from the patient's bone marrow and use a retrovirus that contained the normal ADA gene to insert the gene into the cells. As usual, the retrovirus would be disabled so that it was not infectious. The modified stem cells would then be transfused back into the patient, where they could make unpoisoned blood cells. But results of experimental trials on animals using the stem cell technique did not look promising. Instead, the NIH eventually agreed to a refined technique, whereby the virus was used to insert the gene into white blood cells directly, rather than into bone marrow. In 1990, the procedure, the world's first approved gene therapy human trial was performed on a four-year-old girl called Ashanti DeSilva. The

method went to plan, and, over the next few weeks, Ashanti's parents were convinced that it had improved their daughter's condition. But the scientific outcome of any gene therapy was inconclusive because it was agreed that Ashanti's routine drug therapy, involving treatment with the missing enzyme, should continue. This masked any effects of the new gene. The first official gene therapy trial was, strictly, a test of the safety of the virus-vector technique and no more.

The Pitfalls of Virus-Vectors

The next trial would demonstrate that problems remained in the use of viruses as vectors for delivering genes. In 1999, a similar technique was used to treat a similar genetic disorder. This time the disease was a deficiency in an enzyme called ornithine transcarbamylase (OTC). It caused a surge in toxic ammonia, and, again, most children with the defective gene did not survive into adulthood. Two American paediatricians, Mark Batshaw and James Wilson, began a trial, this time using a common cold virus as the vector. The patient, 18-year-old Jesse Gelsinger, had a massive immune reaction to the treatment, possibly triggered by prior exposure to the common cold. It resulted in his death. An investigation concluded that, for a number of reasons, the trial had not been completed according to proper protocol. The incident was a serious setback in the progress of gene therapy research.

New Developments

The turn of the new millennium saw new viruses being used for targeted gene therapy: viruses that do not cause the kind of immune response that killed Jesse Gelsinger. In 2003, China became the first country to approve a product for gene therapy that could be released for clinical use. It used a virus that contained a gene that shrinks cancerous tumours. By then, entirely new techniques were being added to the gene therapy toolkit.

In 2002, scientists found a way of avoiding viruses altogether by packaging genes inside a protein capsule, and enveloping the lot inside fatty globules called liposomes. The liposomes were small enough to penetrate cells, and even their nuclei. Experiments using "knock-out" mice with cystic fibrosis showed that the therapy worked there. But in trials with human patients, it was found that so few cells are corrected that the treatment is too inefficient for use, even though liposome technology shows great promise for gene therapy as a whole. Meanwhile, other approaches for treating cystic fibrosis have abandoned the idea of trying to administer a functional gene. Cystic fibrosis is caused when cells are unable to make a membrane protein. This leaves tissue lining, especially in the lungs, clogged with thick mucus (see chapter 3). The new approach involves developing drugs that can target the protein-making machinery of the cells directly so that they make working protein.

Other approaches have also avoided introducing new genes. Antisense therapy, for instance, works by disabling harmful defective genes. A strand of nucleic acid (DNA or RNA), called "antisense", is constructed with a base sequence that is complementary to that of the defective gene's messenger RNA. Inside a cell, this messenger is naturally made as a copy of a gene when the gene is being used to build a protein (see chapter 4). By binding to the messenger RNA, an antisense molecule stops this from happening, so the gene fails to be expressed. Antisense therapy has proven useful in stopping some kinds of cancer-causing genes, but also may be effective in treating conditions such as asthma and muscular dystrophy.

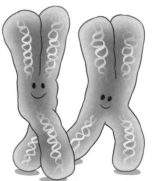

Antisense therapy prevents harmful genes from becoming active, disabling them and stopping them from being expressed.

How We Can Manipulate Genes

Targeted Gene Therapy

Less than a decade ago, the ultimate ambition for gene therapy seemed like an impossible dream: to change the information in genes in a precisely targeted way so that, for instance, the defective cystic fibrosis gene could be corrected into a healthy one. Traditionally, gene therapy had relied on moving genes or blocking them, not on editing them. But a discovery was made in 2010 that brought genome editing closer to reality.

French biologists Philippe Horvath and Rodolphe Barrangou described a previously unknown genetic system found in microbes that helped bacteria destroy invading viruses. Special kinds of viruses called phages specifically attack bacteria, but bacteria had a way to fight back. They carried repetitive base sequence copies of their attacker's genome. These sequences are collectively called CRISPR (standing for "clustered regularly interspaced short palindromic repeats"). Their presence meant they could produce an RNA message to lock onto the virus's DNA, much like how antisense technology can target genes. Once located, the bacteria then destroyed the viral target with a protein called Cas9, which cuts the viral DNA, stopping it from functioning.

Editing Genomes

A year later, American biologist Jennifer Doudna and French biologist Emmanuelle Charpentier worked to take the bacteria's defensive CRISPR-Cas9 system to an exciting new level: to edit genes in human cells. The "CRISPR" part could be fashioned so that it recognizes a defective human gene rather than a virus, while the Cas9-part would get the gene open to deactivate it. That alone would be useful in stopping the expression of a defective gene, but the system can go a step further. By mixing the CRISPR-Cas9 system with DNA that can correct the base sequence of the defective gene, the target cell will incorporate this DNA into its gene when it tries to repair the damage. The entire gene-editing therapy therefore consists of the CRISPR-Cas9 components mixed with corrective DNA, all bound up in a vector, such as a virus or liposome. In 2017 the CRISPR-Cas9 system was used to treat muscular dystrophy in "knock-out" mice, but many trials involving humans are planned, some of which may help treat disorders that are not generally considered to be within the remit of traditional gene therapy. For instance, gene-editing systems are being used to disable genes in blood cells that produce certain kinds of cell-surface proteins. These proteins are used by viruses to gain entry. If the cells fail to produce proteins for binding to, say, HIV, then they will be resistant to attack.

Meanwhile, developments in China have gone a step further, using a system that can edit one or two bases without recourse to CRISPR-Cas9. Using human embryos, scientists reported an almost 25 per cent success rate in correcting the gene for beta-thalassaemia in the embryo cells.

US biologist Jennifer Doudna (above) developed the CRISPR-Cas9 system with colleague Emmanuelle Charpentier. The technique has been used in research into conditions such as cystic fibrosis and Huntington's disease.

Resurrecting DNA

The bodies of living things decay after death, but under some circumstances, their DNA can persist for longer. Some of the best-preserved DNA gives clues about the relationships of extinct species. Modern technology might even offer the promise of resurrecting them.

Like most complex molecules that make up a living body, DNA is fragile. In life, cells are packed with systems that buffer genes from the harsh realities of the outside world. Enzymes and protective proteins keep DNA replicating, producing protein and safely packaged when not in use. But after death these systems quickly disintegrate. Decomposition sets in and complex molecules break down. DNA literally rots.

Only special circumstances can preserve it, and, even then, the vagaries of time mean it never lasts forever. Just as animals and plants that died millions of years ago need luck to become fossilized in rock, ancient DNA is preserved by chance. And the chances are that it will only survive piecemeal. The longer it is left, the more degraded it becomes, so that only fragments remain. DNA can survive remarkably well in specially preserved museum specimens; in nature, it can survive in places that are perpetually cold or frozen, or where oxygen levels are so low that decomposers are hindered. Generally, samples that are more than a million and a half years old lack any intact DNA at all. This means that it is impossible to extract DNA from the remains of most prehistoric organisms. Realistically, conventional techniques, such as gene sequencing, give meaningful results with samples that are only a fraction of this age. But some scientists are exploring ways of improving the technology used to study ancient DNA. And a few are even daring to suggest that ancient DNA could be used to resurrect species that are now extinct.

Nuclear Transfer

The first so-called de-extinct effort began when an initiative led by a biotechnology company began

Can genetic manipulation lead to the reappearance of species that have been extinct for millions of years?

Somatic nuclear transfer is a way of "reprogramming" DNA of body cells so that it is rejuvenated when inserted into an egg. The egg has been stripped of its native genes, so only the donor genes are used to build an animal. The technology was used to produce Dolly the sheep (above) and also in an attempt to "de-extinct" the Pyrenean Ibex.

work on resurrecting the Pyrenean Ibex, a species of European wild goat. The animal had declined since the 19th century due to the combined effects of hunting and competition from other grazing animals. The last of its kind, a female named Celia, was found dead in the wild in 2000. The attempted resurrection used a technique that had been made famous four years earlier, when Dolly the sheep became the first mammal to be cloned from body cells at the Roslin Institute in Scotland.

The technique to produce Dolly was somatic nuclear transfer. Somatic, or body, cells were taken from the udders of a donor sheep. Their nuclei, complete with genes, were inserted into the egg cells from a second sheep, which had its own nuclei removed. Each "empty" egg provided the right conditions to encourage the nucleus to divide and produce an embryo. Dolly the sheep was born in 1996. She had the same genes as the original udder cells, meaning that she was an exact genetic clone of this donor sheep.

In a similar way, body cells from Celia, the last Pyrenean Ibex, were used as donors. Their nuclei were inserted into surrogate eggs stripped of their own genes. In this case, the eggs used came from a common goat. But the procedure was disappointingly ineffective. Of 285 ibex embryos created in this way, almost all died during development. One grew to full term, but it died a few minutes after birth from a lung defect. The exact reasons why the procedure failed are not clear. But the principle of somatic nuclear transfer remains a possible way of helping with efforts to use intensive breeding programmes to conserve endangered species.

One of the last living quaggas (above) was photographed at London Zoo in 1870. A small piece of quagga DNA was sequenced in 1984, and attempts have been made to resurrect this subspecies by selectively breeding the closely related plains zebra.

De-extinction by Genome Editing

For species that went extinct long ago, the problems of resurrection are compounded by the fact that the DNA has degraded over time. Even in comparatively well-preserved prehistoric animals, such as woolly mammoths found in Siberian ice and snow, the genome has become so fragmented that current technologies would not allow it to be pieced back together. Without the complete genome, resurrection of a "pure" mammoth would be impossible, but scientists are exploring other possibilities. Some support the idea of using gene-editing technology, such as CRISPR-Cas9 (described in the previous section), to harvest mammoth genes and introduce them into the cells of elephants, the mammoth's closest living relatives. The result would be a mammoth–elephant hybrid, but a surrogate mother would be needed to carry the foetus.

Other attempts at resurrection take the less invasive approach of artificial selection of living species. The quagga was a race of plains zebra from South Africa. The rear half of its body was solid brown in colour, with zebra-like stripes restricted to its head and neck. It looked so different that until quite recently it was regarded as a completely different species. The quagga was wiped out by hunting, with the last one dying in captivity in 1883. Specimens exist in museums, and, in 1984, the quagga became the first extinct animal to have its DNA sequenced. By comparing museum specimens with living plains zebras, biologists have identified genes that are identical. Since then, a programme of selective breeding has been exaggerating quagga-type characteristics. As a result, in recent years, quagga-like zebras have been produced in increasing numbers.

Some scientists think that de-extinction projects are ethically wrong, and that those resources should be diverted to preventing extinction rather than correcting it. They argue that habitats today may no longer be able to support extinct species and that, in any case, hybrids or products or selective breeding are not *bona fide* resurrections.

How We Can Manipulate Genes

Jurassic Resurrection?

In the Jurassic Park series of movies, based on the original novel by Michael Crichton, scientists were able to recover fragments of dinosaur DNA. These fragments were preserved inside the fossilized remains of mosquitoes, trapped in amber after feeding on the prehistoric animals. By combining dinosaur DNA with DNA from existing reptiles, scientists were able to resurrect multiple species of dinosaur and other ancient reptiles and even genetically create whole new ones. While this makes for good cinema viewing, it remains more than a little far-fetched, as any DNA from more than 66 million years ago would by now be too badly degraded to be used in this way.

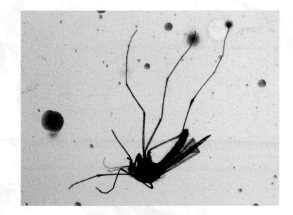

The fossilized remains of a mosquito trapped in amber. The insect would have been trapped by the liquid sap that oozed out of a tree, which then solidified and turned into rock-like amber.

The sight of dinosaurs walking around the modern countryside remains a distant hope of science fiction novels and movies.

Index

Picture Credits

Shutterstock: 6 Double Brain, 8 Kateryna Kon, 10 Sheli Jensen, 12 MidoSemsem, 16m S-F, 16b Egoreichenkov Evgenii, 20 Tewan Banditrukkanka, 22 Panaiotidi, 28, 42, 48t Rattiya Thongdumhyu, 29 Designua, 30 Dr. Norbert Lange, 35l Shutova Elena, 35r ESB Professional, 36 DenisNata, 37 Alila Medical Media, 45 Iren Moroz, 47 somersault1824, 50 Alila Medical Media, 58 Tefi, 78 alybaba, 79 Foxyliam, 82 In The Light Photography, 86 Roblan, 87 Aldona Griskeviciene, 89 Tomasz Klejdysz, 91 Everett - Art, 92 solar22, 94, 102tr Eric Isselee, 98 Taiga, 100bl Mironmax Studio, 100bc L Julia, 100br Aleksandra Voinova, 102bl anyaivanova, 102r kontur-vid, 103l milka-kotka, 103r fotogiunta, 104 Axel Bueckert, 106 XiXinXing, 115 AuntSpray, 119t OZMedia, 119c gcpics, 119b Sebastian Kaulitzki, 124 vangelis aragiannis, 125 Everett Historical, 135 lynea, 136 Mirko Graul, 138t Thorsten Spoerlein, 138b Cristian Mihai, 142t Toni Genes, 142c Dani Vincek, 145t Tono Balaguer ,145b Hayk_Shalunts, 156 science photo, 157 extender_01, 160 Heiti Paves, 162 Mauricio Graiki, 164 Dora Zett, 166 Stockr, 167 oorka, 170, 171 Soleil Nordic, 173 Oleksandr Berezko, 175 Hanjo Hellmann, 177l plenoy m, 177r toeytoey, 179 Elnur, 185t sruilk, 185b Herschel Hoffmeyer

Others: 15 Rocky Mountain Laboratories, NIAID, NIH, 23bl National Cancer Institute, 23br Marc Lieberman/PLOS, 108 Christoph Bock, Max Planck Institute for Informatics/Creative Commons, 142 Evgeny Mavrodiev, Florida Museum of Natural History/National Science Foundation, 148 NOAA, 150 Universidad CES/Creative Commons, 152 Staff Sgt Eric T. Sheler/USAF. 153 Nephron/Creative Commons, 154 National Library of Medicine, 158 National Institutes of Health, 159 Public Library of Science, 165 Bluemoose/Creative Commons, 174 Sam Ogden/Whitehead Institute, 176 US Government, 181 The Royal Society/Creative Commons, 183 Toni Barros/Creative Commons, 184 Frederick York

Illustration Credit

Daniel Limon (Beehive Illustration)

Every effort has been made to contact copyright holders. The publishers will be pleased to make good any omissions or rectify any mistakes brought to their attention at the earliest opportunity.

Publishing Director *Trevor Davies*
Production Controller *Grace O'Byrne*

For Tall Tree Ltd
Editors *Rob Colson and Jon Richards*
Designers *Gary Hyde and Ben Ruocco*
Proofreader *Cynthia Colson*
Index *Helen Peters*